Allan Seale's
Garden Companion to

NATIVE PLANTS

Banksia robur

Stirling Ranges, WA – home to some of Australia's most spectacular native plants.

Allan Seale's
Garden Companion to

NATIVE PLANTS

REED

Reed Books Pty. Ltd
2 Aquatic Drive,
Frenchs Forest N.S.W. 2086

First published 1988

National Library of Australia

Cataloguing-in-Publication Data

Seale, Allan.
Garden Companion to Australian Native Plants

Includes index.
ISBN 0 7301 0187 8.

1. Wild flowers — Australia — Identification.
2. Wild flower gardening — Australia.
I. Title.

635.9'676'0994

Designed by Robert Taylor
Edited James Young
Photographs by Ed Ramsay, Tony Rodd and Allan Seale

Typeset in New Zealand by Saba Graphics Ltd

Printed by Koon Wah Printing Pte Ltd

CONTENTS

INTRODUCTION

The aim of this book is to give readers greater pleasure from Australia's great wealth of unique plant life. It will help them know and successfully grow our native plants, thus preserving them. By knowing a plant better makes bushland or native plant reserve rambles so much more enjoyable. To grow them is extremely satisfying, whether they are in a garden designed especially for natives or combined with exotics, as in fact our plants are by garden lovers overseas.

Identification

Naturally the best identification is from visual illustration. Each of the 400 or so plants dealt with in this book is accompanied by a clear, colour illustration, plus details of leaf type; flower size; flowering time; and area. Where there are differently named plants that resemble each other, for example, boronia, eriostemon and crowea, or bottlebrush (Callistemon) and melaleuca, the simple differences that separate them are explained.

Names

Common names are first choice of many plant lovers but unfortunately these names vary from district to district. In fact different plants often have the same common name. We also have a large percentage of our native flora only known by their botanical name, including popular ones like boronia, grevillea and hakea. As this book is intended as a general reference, the course we have taken is to list plants alphabetically under their botanical name accompanied by their common names. More generally accepted names are cross referenced throughout the book.

For the more obscure or variable names there is the common name index to fall back on, and the illustrations are a good aid for identification for less specialised plant lovers.

Choice

Some of our native flora is most adaptable and easy to grow in most areas. Other plants have been conditioned over millions of years to suit the extremes that exist in different areas, like dry summers with winter rainfall, or the reverse; acid or alkaline soils; mountains or sea coast. Some plants from these extreme areas still have surprising adaptability while others offer a challenge, which some of us do enjoy.

When planning your plantings, it is a good idea to choose the most adaptable plants at least for the garden's general framework. If you enjoy challenges or growing something different, use plants where failure or a short life will not seriously affect the overall landscape.

Suitability

To help you choose wisely, this book gives each plant's area of origin, which apart from interest, gives a better understanding of ideal growing conditions. Included are: width as well as height to which each plant normally grows; type of soil preferred; climate, including frost tolerance, etc.; most suitable aspect; water needs and drought tolerance; type of feeding; pruning; propagation; and whether the plant is prone to any particular problem.

It is hoped that you enjoy learning about and especially growing, our native flora, and that its company brings you much pleasure and satisfaction.

Like stepping stones, sawn log sections add interest and define accepted pathways in bushland gardens.

Planning a Native Garden

Whether you are planning an expansive bushland garden or growing some of your favourite plants in just a few square metres of garden, it is essential to plan before planting. Otherwise it is most likely that taller or quicker growers will deprive smaller treasured plants of sunlight; or else they will completely smother them. If the garden is more or less north facing, it is easy to avoid this problem by planting the tallest growing to the back. If the garden is to be viewed from the south, then you will need to be a little more subtle and keep the tallest plants to either side near the front, or even to a centre island, so that all of the garden cannot be seen from one point.

Then there are the aesthetic aspects of the garden. Whether small or large it is going to look much more interesting and authentic with some background planting to screen boundary fences, intrusive buildings or other aspects out of character with bushland.

If the area is small it will gain an illusion of greater space if soft greys or subdued grey-greens are used in the more distant points, and definite colours kept towards the foreground. Even though the average width given for the background plants may suggest that the available space is being squandered, appreciate that with good initial training and occasional pruning later, most apparently obese giants can be slimmed down to within about a half metre of the fence or even less.

This basically involves fastening side branches back to the boundary and tip pruning or cutting back all forward growth. If started early, most plants can be trained down to any shape or kept within reasonable size.

Another little point worth remembering in the training of plants is that if you need to equalise uneven leading branches on a shrub or tree, you can do this by temporarily staking the retarded one for a few months in an upright position and the more vigorous growth more towards the horizontal, preferably with its tip bent downwards.

Paths can be planned to meander, not only for interest or to create an impression of greater distance but to give different aspects to the foreground plantings especially.

Will the Garden Grow What You Want?

Before getting down to much detail of the plan, consider soil, water and the way your soil retains it. Beds may need raising to improve drainage or sumps may need building to retain water.

Undoubtedly you can go ahead and plant without any changes to the site, if you concentrate on natives indigenous to your particular area, however, some adjustment may be needed if you want to enjoy a wide range of plants from different parts of the continent. For example, plant lovers in the Perth area and further north and to some extent from areas of SA and the inland will need to improve the water holding capacity of soil during summer if they wish to grow plants from the east coast.

East coasters will need to overcome the wet summer conditions that often prevail if they are going to successfully grow plants from dry summer areas. The same may apply to some plants from more moderate climates such as *Dryandras* and particularly *Verticordias* from south WA.

What Do They Need?

Although quick drainage and frequently dry soil is advocated for most plants from dry summer climates, it seems that they still like to get their roots down to a damp subsoil.

Australian native plants can be effectively blended into any type of landscape. Lawn grass is the only non-Australian component in this garden.

Yellow Buttons (*Helichrysum apiculatum*), Native Violet and Grevillea Royal Mantle carpet attractively below pendulous Grey Melaleuca (*M. incana* at left). An unusually tall Native Hibiscus (*Alyogne heugelii*) grows above Grevillea Robyn Gordon and a bottle brush, which both bring birds close to the windows.

Compact pink flowered *Pimelea ferruginea* surrounded by *Grevillea biternata.*

Achieving this dry topsoil, damp subsoil situation is not as difficult as it might sound.

Successful east coast grower Sid Cardwell achieved it by spreading a 10–15 cm layer of coarse river sand and a dash of compost on top of the reasonably moisture retentive loam. This added soil was retained in suitable sized beds with timber or masonary edging. The original level paths between beds than acted as drains.

John Hunt in his book *Creating an Australian Garden* demonstrates how he successfully used extensive mounding for his most effectively landscaped large native garden. Rain or watering which to a large extent would otherwise run off the mounds was trapped by broad gutters with closed ends between or surrounding mounded areas. Going a step further, ditches were dug into the subsoil below these gutters to store a longer lasting sump of water for diffusion to the surrounding subsoil. From this, water is supplied to plant roots by capillarity, and at the same time these roots are encouraged to penetrate deeply.

With a little planning, some stepping stones and in sticky soil areas, a few centimetres of clinker ash, these broad gutters could become paths surrounded by carpets of Scleranthus, Ajuga, Isotomas and other moisture-loving natives.

Trees

Any garden needs at least one or two trees to give it character. These need not be large. Towards the north, west or centre of the garden choose light foliaged types like scribbly gums, mulga type wattles, *Albizzia, Agonis juniperina, Grevillea* 'Sandra Gordon', etc., as all give only dappled shade. Also consider that plants can be set out under them so that they receive full benefit from the lower angled sun during winter and some protection from hot summer sun when the dappled shadows will be more directly below the trees. Keep dense foliaged trees or large shrubs to the south of the garden.

Scribbly Gums (*Eucalyptus haemastoma*) provide dappled shade and typical bushland character.

Year-Round Appearance

It is more usually the beauty of flower that determines the choice of plants but think about foliage and foliage combinations that will keep the native garden looking interesting throughout the year. There are plants grown for foliage alone including *Hypocalymma* 'Golden Veil', *Melaleuca bracteata* 'Golden Gem' and others, but just by thoughtful combination of different greens and greys the overall effect can be made more interesting.

Combinations of plants with different leaf textures and particularly of different forms can add a great deal of interest. In fact a group of well-placed plants with contrasting form, say a conical type towards the rear of a smaller dome-shaped one, then balanced by a still smaller decumbent grower, can in itself provide a feature. The combinations that can be used to create interest are unending, apart from the wonderful bonus of their flowers.

On a smaller scale a raised rock garden can cater for different degrees of drainage as the higher levels would normally be the drier ones. It also allows for variations of soil in different pockets and for different aspects, from warm sunny pockets on the north and comparatively cooler and more shaded ones on the south side.

There are so many ways to create a natural bushland atmosphere in the native garden. Logs can be used to retain soil. If possible avoiding placing them parallel to obvious boundaries, buildings, etc. A weathered rock or two may be allowed to protrude from the edge of a mound or set near the trunk of a small tree.

Woodchip mulch makes a natural looking soil cover. It also helps to bind mounded soil, acts as a buffer against heavy rain that would otherwise puddle and ruin soil structure, and keeps soil temperature more even. Combine sticks, stones, leaves, etc., and later add a few grassy leaved natives like *Sowerbaea, Patersonia* and small groundcovers in appropriate place to emphasise the bushland look.

Leaves make an effective mulch and weed-retarding cover but when applied thickly on their own tend to shed rather than encourage water penetration. However the latter point can be a help rather than disadvantageous when depending on the capillary watering idea suggested earlier because it minimises evaporation from the soil. Leafmould also gradually feeds plants as it decomposes. It is safest to keep it back a little from the stems, especially those from dry summer climates that are being grown in humid areas such as the east coast.

Tree branches add an authentic bushland touch and are also practical to retain mulch and soil on sloping sites.

Cultivating Native Plants

The Most Important Step — Planting

The way a plant is started makes all the difference to its survival and future performance. Unfortunately their roots are too frequently buried rather than planted!

Understanding Soil

It is not always remembered that plant roots need small amounts of oxygen before they can develop and function. Like water, this oxygen enters through tiny pore spaces between the soil particles. The coarser the soil particles the larger the spaces between them and therefore the easier the entry of oxygen and water. Naturally there is an optimum here because beyond a certain size space the water runs straight through except for the small amount held by organic particles mixed with the soil.

This need for oxygen also explains why many plants deteriorate in overwet soil, because when the spaces between the soil particles are filled with water, oxygen is excluded. The plant can then actually die of thirst because the oxygen-starved roots cannot function sufficiently to take in water, even though they are immersed in it. Then the plant shows much the same symptoms as it would from excessive dryness.

Root performance is also usually poor in clayey loams that have been puddled by heavy rain or watering. Digging, planting or agitating the soil in any way while it is wet or sticky also puddles the soil. In this case the ultrafine clay particles normally bound together into comparatively coarse crumbs,are forced into solution with the water in the soil by the agitation. The resulting mixture is mud, an airless mixture, perhaps suitable for making bricks or pottery but not for growing plants. Therefore plant only when the soil is just damp, never when sticky.

Also, after planting in heavy soil, even before initial watering, better root development will be assured if a mulch of leafmould or woodchip is placed over the root area to prevent surface puddling. The latter results in crusting of the surface area which retards aeration and even entry of water. Even in light sandy soils, the surface mulching helps to retain the organic particles in the soil which would otherwise wash out during heavy rain or watering and deplete soil structure.

Acid or Alkaline Soil?

Many of our native plants adapt to a fairly wide range of soil acidity or mild alkalinity. The latter is more usually due to relatively large quantities of lime in the soil.

This is a factor which should be considered, especially when growing plants native to other areas of the continent as some of these do not accept extreme changes of soil acidity. Generally speaking, soils in fairly high rainfall areas are relatively acid and those in low rainfall are frequently alkaline due to higher content of lime or other alkaline materials which would otherwise neutralise acidity.

The reason for this is that frequent heavy rain leaches lime from the soil. However, there are exceptions, such as soils with high iron content in low rainfall areas where any free lime is formed into insoluble compounds by the excess of iron, or where there are heavy limestone deposits in high rainfall areas. In this case it is possible that the river water is limey, even though most soil in the district is neutral to acid. Apart from possible intolerance of some plants to excess of acid or lime, the important factor is that many essential plant elements become insoluble in soil that is excessively acid or alkaline.

Wood chip mulch preserves soil structure and inhibits weeds while mixed ground cover plantings are establishing. (*Brachycombe* in foreground).

In this insoluble state these elements cannot be absorbed by plant roots. Growth then becomes stagnant or in some cases deficiency symptoms occur.

How does one determine whether a soil is acid or alkaline? There are simple to use soil-testing kits that give this answer fairly accurately. Rough indicators are hydrangeas which usually produce blue flowers in acid soils and pink ones in neutral to limey soils. However, it should be remembered that some part of a garden can become limey, such as near brick work where there have been considerable mortar droppings or where builders have mixed these materials. Here hydrangeas may continue producing pink flowers for many years, even though the soil may be quite acid in other parts of the property. Soil close to new concrete may also remain limey for some time. Healthy growth of bracken, or azaleas and rhododendrons growing without use of soil acidifiers, will also suggest moderately acid soil.

More accurately, the degree of soil acidity or alkalinity is measured by the pH scale, which ranges from 1 to 14 with pH 7 representing neutral. The lower the figure below 7, the stronger the acidity and the higher above 7, the more alkaline.

The type of soil-testing kits available for home garden use from comprehensive garden stores containt an indicator liquid which changes colour when applied to a soil sample. This colour change is then matched with the colour chart provided and the corresponding pH figure noted.

Below pH 4, with the exception of iron and magnesium, the essential elements become almost insoluble and therefore plants make little if any appreciable growth. pH4–5.5 is regarded as strongly acid, pH 5.5–6.5 mildly acid and desirable for a wide range of plants, because most of the essential growth elements are then freely available. pH 7–7.5 can be regarded as slightly alkaline or limey and suitable to many plants from lower rainfall areas. In this range all necessary elements remain reasonably available except iron and to a lesser extent manganese.

Without iron plants cannot make chlorophyll, therefore if limyness reaches or exceeds this point, the leaves may begin to yellow. Because of the insolubility of iron and immobility within the plant, the yellowing will also show first on the new growth.

From these remarks it becomes obvious that excess acidity can be reduced by adding lime to the soil. The amount needed will vary according to soil type. Only garden lime, or dolomite should be used. The former is also referred to as agricultural lime, ground limestone or carbonate of lime.

Changes of soil pH should be made gradually, especially when decreasing acidity (raising pH). Apply no more than 300 to 400 g of lime per m^2. and dig it into the soil. In many soils this would raise pH by about 0.5 but leave for about one month after the first application then check again before applying more. Remember in all cases that an excess of acidity or alkalinity can be detrimental, therefore 'hasten slowly'.

Alternatively, in relatively acid east coast soils and Canberra Botanic Gardens, digging limestone chips into the soil has proved successful in establishing natives from limey areas of WA and SA.

Root-Bound Means Retarded

If planted in a root-bound condition, i.e., with the plant roots wound around and matted against the container wall, they invariably stay that way, with few roots, if any, venturing out into the soil.

Even though the plant may appear to make good top growth after planting, it eventually becomes retarded and dies suddenly, because of its restricted root system, usually during hot conditions due to water stress. With many shrubs, it is possible to unravel or trim off restricted roots if planting during dormancy. This encourages a new root system but only a few Australian natives will accept such drastic disturbance.

This shrub died during dry conditions about six years after planting. Even though its growth had nearly doubled in that time, examination showed that roots had hardly ventured from the original root ball. Similar root restriction sometimes occurs when soil in the planting site is particularly heavy or much poorer than in the plants container.

Illustration of typical shrub root growth (in this case *Prostanthera*) obtained by burying a sheet of fibro-cement just below the surface 25 to 30 cm out beyond the foliage line. It demonstrates that contrary to common belief, feeding roots do extend well beyond foliage spread and that the plant is easily injured or debilitated by any but very shallow cultivation. It also shows how roots will congregate below nearby flat rocks to take up the moisture from condensation and capillary movement.

The Answer

Either transplant while the seedling or cutting is young or do not buy plants in a root bound condition. If roots are winding and matting around towards the base of the pot, gently unwind and tease them out. Cut off any broken badly distorted areas or tight woody masses. With most plants you should get away with this if the top growth is cut back to compensate for the root loss. It is better to risk losing the plant than have it a permanent cripple.

Avoid buying very advanced native plants. Even though they may not appear root-bound it can be difficult to determine their root condition prior to the time of replanting into their present container. In any case, within a year they are often bypassed by others that were half their size at planting time.

Soil Acceptance

A point that many gardeners do not realise is that even though a plant may not be root-bound at planting time, if the soil in its container is in a much better condition than that in the planting area, its roots may refuse to leave the root ball. They then continue to encircle the soil ball and in effect the plant becomes root-bound.

Avoid this problem by making a hole wide enough (at least twice the plant container width, preferably more) so that the 'improved soil' can be filled back around the root ball.

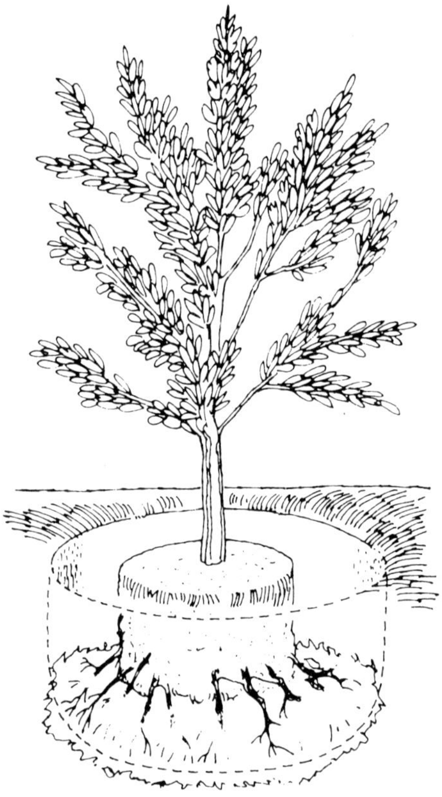

Soils in containers are often better than the soil in which the plant will be finally positioned. Roots can remain root-bound in the better soil. Avoid this by digging a hole wide enough to surround the plant with improved soil.

Improved Soil

If soil in the planting area is rather clayey, very sandy, seems in generally poor condition or you are in doubt about this, mix with the backfill removed from the hole, up to one-third leafmould, or in the case of tough clay-loam that does not crumble easily, about one-third leafmould and one-third coarse river sand. Peatmoss, previously moistened evenly, could be substituted for leafmould in quick drying sand or in other cases where the recommendation is for 'moist soil'.

Don't Bury the Poor Plant

Planting too deeply is often the cause of complete failure or, on some occasions, a plant just sitting and refusing to make even one new leaf in a year! The excess soil over the surface of the plant's soil ball is sufficient to retard entry of oxygen and causes suffocation of roots. This is why the general recommendation is for a wide hole but *not deeper than the soil ball.*

Depth can be deceptive, therefore it is better to check it while the plant is still in the container. In fact when planting natives that need good drainage in heavy soils it is usually better to have the soil ball 1 cm or so higher than that of the surrounding soil and gently slope up to it with improved soil.

Cover the soil ball with some decomposing leafmould by all means but not with soil.

Step by Step Planting

1. *Have soil just damp,* never wet and sticky or powder dry.

2. Dig the planting hole *at least twice as wide* as the plant container but *do not have it deeper.* Alternatively if the base has been deeply dug to improve it, then tread it down well before planting. Do not dig into clay subsoil.

3. *Water plant* an hour or two before planting, allowing plenty of time for water to drain completely. Remove plant from container by inverting with hand over soil ball and giving a few sharp taps on edge of container; if in a plastic bag, cut away the plastic.

4. Position plant so that *container soil level is same as surrounding soil* or slightly above. Adjust this by firmly mounding soil in base of hole.

5. Carefully *tease out roots;* if entwined at base of container, spread and cover. See under heading: Root-Bound Means Retarded.

6. Gradually *fill in with improved soil* (see heading: Improved Soil), firming evenly around plant (with hands not feet) to fill any air pockets.

7. Finish by fashioning surplus soil into a water-holding ridge just in from the edge of the planting hole. Cover this saucer-shaped surround with a few centimetres of either leafmould, dry grass (not grass clippings unless partly decomposed) or woodchip, then give a good gentle soaking.

Note: This surface mulch applied before watering helps to keep the soil surface more open. With any of the few dry area plants subject to stem rot, it can later be raked back a few centimetres.

Watering

Even with dry climate plants it will be necessary to keep the soil damp until they are established — at least for the first six weeks. However, with all but the very few exceptions that need continuously moist soil, continue to give good soakings but less frequently. Initially allow no more than the top centimetre of soil surface to dry between soakings and then after a couple of weeks allow the dryness to extend a little deeper so that roots are encouraged to penetrate deeper. In this way the plant becomes more accustomed to dryness.

Staking

Forget about it except in very windy aspects where some compromise may be needed. Yes, it has been proved that plants staked in an upright position grow quicker but they do not develop wind resistance and invariably topple during the first strong wind after the stake has rotted.

If you have to plant something tall and top heavy it will be difficult for new roots to establish if the soil ball is frequently rocking. In this case place a firm stake 10–15 cm out from either side. Then fasten the tying material firmly to the stakes but loosely crossing the plant in a figure eight so that it reduces but does not eliminate root movement.

Far better still, copy what nature does in windy areas. Either start with a very small plant that will grow away from the wind or with a plant that has its stem at a fairly steep angle away from it so that either the main stem or some of the lower branches are actually touching the ground and the plant becomes completely self-bracing. The growing tip and subsequent growth will then become upright. This is the way Geraldton wax, many of the *Melaleuca, Grevillea, Hakea* and other plants grow naturally in exposed areas.

Nurserymen frequently need to stake to protect and prevent entanglement of plants grown or handled en masse, but it pays to carefully remove these stakes when planting. Even if the plant is reasonably well-grown, either let it sprawl or cut it back to encourage low branching.

Geraldton Wax should be grown without support from stakes. Lower branches will then brace it against strong winds.

Introducing New Plants

It is not unusual for apparently healthy plants to give disappointing results when added to gardens where most plantings are already well established. This is usually because the newcomers alone receive frequent watering, mulching and in some cases feeding.

Roots of surrounding plants then invariably move in to the specially favoured positions for their share of the handouts and within a surprisingly short time the poor little newcomers are fighting against overwhelming opposition while trying to establish their own feeding system.

The answer is to eliminate this competition by watering the entire area, or if this is going to upset your watering programme for nearby plantings, then minimise the newcomer's problems by instigating a few decoys such as leaving a hose trickling occasionally between the new and old or on the latter's opposite side — perhaps adding a little fertiliser to the decoy areas.

The same principle applies when feeding. There are times when it would pay to throw a few handfuls over your neighbour's fences to deter their tree and shrub roots from poaching on handouts intended for your plants.

Feeding

This is a controversial subject because it is often stated that Australian native plants should not be fed while some growers claim outstanding results from feeding. It would be reasonable to say that the correct practice depends a great deal on the plant species, the possible growth activity at time of application, the moisture content of the soil and the type of soil.

Phosphorus, which with nitrogen and potash is a major component of basic complete (dry mix or sprinkle-on type) fertilisers, is often the culprit when adverse results occur due to feeding because it is the major plant-feeding element most deficient in Australian soils. Some plants, especially *Grevilleas, Hakeas* and others of the Proteaceae family have specialised roots which can absorb and store comparatively minute traces of this element and therefore what would be regarded as normal applications for exotic plants is toxic to these natives. Rutaceae, Epacridaceae and Goodeniaceae are among the plant families rather sensitive to this fertiliser.

One interesting point that could account for some of the contradictory results from this fertiliser is that in very acid soils, such as Hawkesbury sandstone, other types of sandstone or soils where lack of fires has allowed years of leafmould decomposition, phosphorus is almost insoluble and therefore unavailable to the plants. This erratic element, however, is highly soluble in slightly acid and neutral soil, but loses a great deal of solubility in slight alkalinity then gets it back again with excess alkalinity or limeyness.

Sulphate of ammonia, the ingredient generally used to supply the nitrogen in basic complete fertilisers can give a great boost to most of the Myrtaceae, Proteaceae, Rutaceae and others if used discriminately. However, it can have a retarding effect on many of the Papilionaceae family (the 'pea' flowers) because they have the ability to use free nitrogen absorbed from the air with the aid of different types of symbiotic bacteria present in clusters of small nodules on their roots.

This is why apart from its direct burning effect when lodging on wet foliage, sulphate of ammonia will minimise clover in lawns or invasive native peas like tiny purple-flowered *Glycine clandestina* which germinates freely, rapidly entwines and soon smothers any plants within its long reach. Inversely, superphosphate or other fertilisers containing mainly phosphorus will usually cause these pea-type plants to predominate.

Burning Danger

Another reason for being cautious in applying fertilisers, in some ways the most important one, is that damage to the roots can result from what is commonly called burning. This damage occurs when all the soluble salts from the fertiliser ingredients go into solution with the soil moisture and strength of that solution is greater than that of the solution in the root cells. Instead of the cells absorbing the nutrient salts and passing them throughout the plants cells, the action is reversed and the stronger solution draws liquid from the weaker one (exosmosis). This causes the plants cells to break down.

It is to minimise this damage that it is suggested that the soil should be moist before applying plant foods and the moisture content should be maintained for usually a few weeks after feeding because as the soil loses moisture, so the concentration of this soil salt solution proportionally increases.

Different plants vary in their tolerance to soil salt concentration. Most plants accept a 6–8 per cent salt solution whereas most of the Ericaceae family accept only about 4 per cent and it can easily be seen that these percentages can increase many times as soil dries out. This is the reason why proportionately small quantities of fertiliser are recommended for pot plants where there is less diffusion into surrounding soil and why it is a good idea to leach soil in containers (give a good soaking) to wash out unused fertilizer, before feeding again. It is also a reason why 'half as much twice as often' is suggested for sandy soils, partly because plant foods leach out of them more rapidly than in heavier soils but also because they dry out much quicker.

Yet another important point is that plants should only be fed when they are capable of or, safer still, beginning to make growth. For most plants this is as flowering finishes or a little after. The main exceptions are some of the WA plants, including some kangaroo paws, which have a period of dormancy during summer and therefore are fed in autumn.

It is a good idea to always use compost, leafmould or other organic material as well as fertiliser as this slowly produces a continuous supply of nutrients in safe quantities over a longer period and maintains the soil in good condition.

How much?

The reference in the text to 'moderate' application, for a general complete fertiliser, would mean about one handful/m^2 of area below the plant's branches and about 0.5 m beyond the branch line, concentrating more on the outer area. This should be raked lightly into the surface soil or mulch then watered well to take it down into the soil.

With young eucalypts and other tall trees or large shrubs where deep root growth is to be encouraged, some enthusiasts prefer to make crowbar holes below outer branches to the depth of subsoil, about 1 m apart and pour the equivalent amount of fertiliser into these.

Where the text says 'if at all' under the heading of Feeding, use about half the amount suggested for 'moderate'. However this is suggested only if one is seeking

Native plants like Honey Grevillea (*G. eriostachya*) attract birds and in turn gain some nutrient from their droppings.

extra performance from these plants as they will survive happily in nature with the nutrients from decomposed leaves, plus chance bird, other animal or insect droppings. The bird droppings will probably be more plentiful around *Grevilleas, Callistemons* and *Melaleuca* (as these attract wattle birds, spine-bills, parrots, etc.) than they would around other plants. The droppings are conveniently deposited at flowering time ready for the new growth that follows!

What Fertiliser?

Previous reference has been made to standard complete fertilisers. There are also 'complete' organic fertilisers which have mainly an organic base — probably blood and bone with the absent major element potassium added in form of potassium chloride or better still, potassium sulphate. Some of the better and usually more expensive complete fertilisers also have a trace element balance added. This is a decided advantage because trace element deficiencies do exist in some areas but generally speaking do not become a problem unless the three major elements nitrogen, phosphorus and potassium (NPK) are used in high concentration, used frequently, or one of them is used alone. This can upset the balance and cause deficiencies of minor or trace elements to become apparent.

Organic Fertilisers

These are based mainly on cotton seed and other vegetable meals. They are slowly broken down by soil organisms into soluble inorganic salts, etc., that plants can absorb. However, this breakdown is usually a comparatively slow process and because nutrient salts become available gradually, the risk of burning is minimised.

Slow Release Fertilisers

These granules are coated with a permeable film that allows nutrient salts through according to soil water content. This happens gradually over a long period, usually feeding the plant for three to six months, as stated on the container. Therefore they are generally safe to use, especially if applied in spring or when the plant's normal growth period begins.

Complete Water Soluble Fertilisers

These are in crystal or powder form, usually dissolved at the rate of about 1 heaped teaspoon to 4–5 litres of water and watered around the plant's root area. At the dilution rates recommended these are safe for most plants if applied during the normal growth season. The only criticism of some popular brands is that when used frequently without other forms of feeding, because of their high nitrogen content they are inclined to produce lush growth at the expense of flower. However, their comparatively low phosphorus content makes them safe for most native plants if not used more than about once every three to four weeks during the growing season and of course in conjunction with leafmould or compost to keep the soil in good condition.

Seaweed Extracts

These are sometimes criticised because their nitrogen: phosphorus: potash (N:P:K) ratio does not conform to what is regarded as acceptable. However, apart from the stated contents of these chemicals, by results they have shown that they either contain substances that act as catalysts to improve the intake or availability of other nutrients and also promote the growth of soil organisms beneficial to plant growth.

For example, they promote the growth of certain protozoa in the soil which can utilise nitrogen from the air, and this nitrogen passes on to the plants as these millions

Kangaroo Paws (*Anigozanthus flavidus* hybrids) stand sentinel by the rock garden at the National Botanic Gardens, Canberra. Here, plants which require limey soil (high pH) are grown in soil mixed with limestone chips.

of microscopic organisms complete their short life cycle. The same could be said for the old fashioned 'liquid manure' brews that are often scoffed at in these days of modern chemical 'know how'.

Animal Manure

All manures contain nitrogen, small amounts of phosphorus, depending on the herbage grazed by the animals and if stable manure, a little potash. Fowl manure is the only one consistently high in phosphorus as well as nitrogen. It should only be used in small quantities — a scattering rather than a layer.

Never apply mulches of animal manure in a fresh state. Apart from the detrimental excesses of ammonia often present they can generate sufficient heat to damage plant roots and in some plants induce stem or crown rot. The same applies to still-green grass clippings or other foliage. Sprinkle it lightly if you care to but before using it for mulch it should be moistened and heaped for a few weeks. It can make good compost when mixed with gum leaves and if you have it, a little animal manure.

Safety Rules

1. Feed only when soil is moist and water well after application. Do not feed if follow up watering is not possible.

2. Do not exceed quantities recommended — it is better to err on the lighter side as more plants die from overfeeding than starvation.

3. Do not feed during seasons when plants do not normally make growth — it is safer to wait until new growth starts.

4. Do not feed recent plantings. Apart from danger of burning new roots, plants lose the incentive to make deep root growth if plant foods are close at hand. Some growers do place slow release fertiliser at the base of planting hole, but it is better to place this deeply near the subsoil to encourage root penetration to this relatively moist area. Make sure soil above it is well firmed before positioning plant.

Pest Control

Here are some suggested controls for pests and diseases listed under Possible Problems in the individual species entries. Non-toxic controls are given where possible and practicable, otherwise the least toxic remedy is suggested. Sometimes an alternative is offered, but only where it is not highly toxic and dangerous to use.

Before waging chemical warfare when a pest problem arises, or becoming despondent about its outcome, remember that there are many natural predators keeping the plant-attacking types in check. Nature maintains fairly tight control on this ecological balance and without too much human interference rarely lets things get out of control.

The cost of keeping this balance is an occasional chewed or mottled leaf. We must bring ourselves to accept this small price because if all plant-attacking insects were killed off, any useful predators surviving the insecticide used would die of starvation. This in turn must affect the insect-eating birds, lizards, etc., and any plant pest surviving could multiply unchecked into plague proportions.

Just to show you how many different predators of common plant pests there are, consider the following:

Caterpillars are parasitised by ichneumon flies and other tiny wasps which deposit their eggs in the caterpillar's bodies. The latter do not die until approaching pupae stage when larvae of the predators emerge but it is a control which halts the cycle. There are also wasps which parasitise cocoons, even the hard woody ones of the gum-tree-stripping cupmoth caterpillar. Insect-eating birds dispose of caterpillars and other plant pests far more rapidly, and often in large numbers.

One ladybird eats fifty to sixty aphids a day. Even that little dragon-like blackish ladybird larvae has a daily diet of twenty to twenty-five aphids. Then there are the tachinid flies, robber flies and many more direct attackers which dispose of insects immediately. These include dragon flies, praying mantises, assassin bugs, both the brown and the graceful green lacewing flies and their ugly little larvae, centipedes, spiders, lizards, which also eat snails eggs, etc., and of course the birds.

However, don't think I'm suggesting in any way that plant attacking insects should be fostered. Squirt off or rub off aphids by all means and look for the culprits where there is caterpillar damage and dispose of them as you choose — there will always be some you have missed for the birds or other predators. What I'm trying to point out is that chemical control, which for too many years has been regarded as a major virtue of gardening, is not always necessary, can upset the natural balance, and incidentally, can do still greater harm to the user.

Care with Chemicals

When chemical sprays are used, always follow label directions carefully, particularly the safety directions. Do not spray on a windy day because no matter how careful you are, you will invariably end up receiving more spray than the plants. Also avoid spraying in the heat of the day as this may result in foliage damage. Always keep chemicals well out of children's reach, preferably in a locked cupboard.

Suggested Controls

Aphids

These can be rubbed off or removed with strong jet of a hose when clustered below

The eggs of a useful predator, the Green Lacewing Fly. These are also found on fence posts or other timber, window panes etc. and are often mistaken for spores of a fungus. Like many predators, the larvae are carnivorous. To minimise the likelihood of the first hatched devouring neighbouring eggs, nature made it necessary for each to descend this difficult hair-like support before gaining normal mobility.

These encrustations shelter 'lerps' (a form of psyllids) on gum leaves. They usually cause browning and shedding of foliage.

buds or new growth. If under foliage they need spray contact. Non-toxic; use soapy water or, better still, garlic spray for all but black aphids. Pyrethrum is even more effective. Relatively low toxicity Malathion, or moderate toxicity Lebaycid or Rogor is probably more effective when pests are under foliage and difficult to contact directly. However, in the writer's experience, when spraying of an aphid-infested large shrub was delayed because the presence of ladybirds and their eggs was noticed, this heavy infestation was gone within 3 days. So give ladybirds a chance where present.

Smaller birds, especially wrens and silvereyes in small flocks, will systematically move from shrub to shrub (or trees) relieving them of aphids and other small insects.

Thrips

These may rasp and can cause greyish mottling on cissus and a number of broad leafed shrubs and cause flower damage, usually during hot dry conditions. They can be deterred by misting foliage with water, better still with one of the soapy garlic sprays. However Malathion, which is mildly toxic, does give more effective control. Remember, however, it is also toxic to bees.

Mites

Two-spotted mite or red spider mite, like thrip, is usually prevalent during hot dry conditions. They are almost too small to detect by naked eye, except perhaps in bright light, even though many hundred may be present on a leaf, usually on underside. They usually cause a dull and lustreless or sand-blasted appearance. The same control as suggested for thrips, particularly frequent misting with water deters them. Chemical controls which do not affect other insects are: dusting with sulphur or spraying with wettable sulphur. Both are non-toxic but can cause foliage burn of some sensitive plants during very hot conditions. Another method is to spray with a miticide such as dicofol (Kelthane) which kills eggs as well as mites, although some mites have resistance to it and some plants are sensitive to the chemical. Where mites are a problem in large scale crops, the introduction of predatory mites has been successful. These are available commercially.

Psyllids

There are a number of different types and some, including the 'lace lerp', shelter under tiny shell- or fan-shape whitish encrustations, 3–5 mm across. These are sometimes found on gum leaves, usually causing them to brown and fall. They are evidently devoured by birds or predatory insects because infestations rarely occur the following year, or at least not in the same area.

Lace lerps can be controlled on small trees where safe spraying is possible by Rogor or Lebaycid with the addition of a wetting agent or about a level tablespoon of white oil per 4– 5 litres of water.

Another form of psyllid causes the blistering of lillypilly (*Syzygium*) leaves. Attack can usually be minimised by spraying when symptoms are first noticed, using the mixture suggested for lace lerps.

Grevillea psyllid can cause stunted and deformed low growth and also shrivelling and fall of flower buds, especially on *Grevillea rosmarinifolia* varieties. This type of psyllid is very mobile, like an elongated white fly or minute cicada with a mealy appearance and may hop away vigorously when disturbed. They feed on the plant sap like a mosquito on blood and it is believed that the saliva they inject in the process is toxic to the plant. It appears to be the same psyllid found on the young growth of maples and other shrubs.

Caterpillars

Caterpillar pests on natives include the callistemon tip bug and tip-rolling caterpillars. The latter particularly will frequently attack young shoots of a number of natives including eriostemons and boronias. The tiny caterpillar frequently webs young tip leaves together as protection then eats out the newer growth forming below.

This may seem devastating but in some ways it can be regarded as a beneficial form of tip pruning which results in bushier and more compact growth. However, the tip is not always killed and when growth continues the chewed or perforated leaves are often left further down the stem.

The best idea perhaps is to squeeze the webbed tip in hope of eliminating its tenant then pinch or cut off the damage(tip prune). This tiny caterpillar is quick moving and will probably back out the back door rapidly and remain suspended on a fine web between tip and ground while you are attempting to open the webbed leaves, therefore one hand under it first is a good idea. Lebaycid or Rogor (each moderately toxic) are sometimes used to break severe infestations.

Leaf-webbing or bag caterpillars form small colonies on flame trees and kurrajongs (*Brachychitons*). Those in reach can be cut off and disposed of. Otherwise most good garden sprayers can be adjusted to a long jet reaching to about 10 m, to facilitate spraying with carbaryl (low to moderate toxicity) plus a wetting agent. Even though it may not directly contact the caterpillars it should poison most of them as they feed.

SAWFLIES

These are revolting greyish black caterpillars 2–4 cm long, with scant hairs.They eject a brownish foam when disturbed and hang clinging together in colonies, usually on gums but often on other trees, including *callistemon*. Birds are usually too discriminating to go near them but they do seem to be parasitised by tiny wasps. They can also be controlled as suggested previously for leaf-webbing or bag caterpillar.

WEBBING CATERPILLARS

These occur on tea trees (*Leptospermum*), some *Melaleuca* and occasionally some *Kunzea*. They are usually brown to grey, smooth, slender caterpillars that harbour in colonies within densely webbed areas. Some envelope a number of branchlets, others only a few leafy twigs with only a few or even just one occupant. They emerge, usually at night or under webbed cover, and strip foliage.

Webbing makes spray contact difficult. It may be possible to dismantle most of the web with a stick then spray with pyrethrum (toxic only to insects and fish). Also spray ground below shrub. Alternatively spray with carbaryl (mildly to moderately toxic) plus a wetting agent which will poison them when they emerge to feed. With very dense foliaged plants like *Melaleuca incanum* 'Velvet Cushion' it is usually necessary to cut away closely webbed areas.

LOOPERS AND OTHER MARAUDING CATERPILLARS

These include the hairless green or brown loopers, so called because they travel with a loop action, stretching and clutching with their front set of legs then pulling the back ones under the front of their body. When disturbed they either clutch the stem with rear legs and project themselves at the same angle as a branch stub or lie along the mid-vein under a leaf. They always take on the same colour as their host plant.

They even pose as the broken margin of the leaf where they have been feeding. On young gum trees, particularly *Eucalyptus citriodora*, one form most effectively poses as a young leaf cleanly eaten down to within 1 or 2 mm of the centre vein.

Once the eye becomes accustomed to their camouflage it is fairly easy to pick them from any but large plants by hand. Otherwise spray thoroughly with pyrethrum which has a residual effect of about one day, or with mildly to moderately toxic carbaryl which lasts for about five days. Both are toxic to bees, therefore avoid spraying flowers or use pyrethrum late in the day when bee activity has ceased.

Dipel, a non-poisonous bacterial culture, kills a wide range of caterpillars if sprayed evenly over foliage and does not affect bees or other insects. However, it does not kill all types of caterpillars and sawflies.

Beetles

A number of similar beetles chew foliage, especially of gum trees. These can be treated as suggested for caterpillars. Their larvae, which are various types of curl or chafer grubs live in the soil and can do considerable damage to roots of plants. They may be particularly troublesome in container-grown plants, perhaps because of absence of predatory soil beetles, etc. Badly infested areas can be treated with fenamiphos granules sold under such names as Lawn Beetle Killer.

DENDROBIUM BEETLES

The adult beetle sometimes frequents ageing *Cymbidium* blooms then turns attention to *Dendrobium*, particularly *Dendrobium falcorostrum, D. kingianum* varieties and *D. speciosum*, when new growth is emerging. It strips fleshy tissue from the tender young leaves, often leaving little more than the leaf veins, then its maggot-like larvae, usually covered in a foamy substance, bores into the pseudobulbs, often causing them to break off where it later emerges or to rot.

The adult, with its main body like a straight-sided shield with rounded base and its neck tapering to a small head with prominent antennae, is a 12–13 cm long, glossy, but dull orange-tan, blotched with black at both ends of the wings.

As they more usually drop and are hard to find or sometimes fly when agitated, if possible quickly place one hand below the leaf they are attacking and grab with the other. However, if they do fly it is usually in a slow circular path, returning to a point near the plant, so poise for a successful catch. When unable to seek out the beetles, preferably early or late in the day, spray plants with carbaryl, concentrating on young foliage.

Borers

There are many types. Those more commonly affecting *Grevillea, Hakeas, Callicoma, Ceratopetalum*, some *Leptospermums* and other shrubs usually do not bore deeply into the stem and are indicated by what appears like a patch of web and sawdust. Another type is a little less obvious, covering its hole and area it has ringbarked with a closely woven web fabric a similar colour to the bark. It is detected by its slightly different texture, slightly greyer colour and by a slight bulbousness.

Control those described by removing the fabric and probing with copper wire or by injecting a few drops of Malathion or methylated spirits. The moth larvae may then emerge to be squashed. The hole should then be puttied to prevent water entry.

Longicorn borers in the adult stage are large slender beetles or giant weevils to 6 cm long, usually brown or various shades of grey, sometimes with whitish or buff marking and antennae as long or sometimes longer than the body. They emerge from late spring into summer, the females laying eggs in cracks of the bark or particularly in points of injury. The larvae, a long, segmented, cream to light brown, club-shaped maggot (legless) tunnels into the soft wood and growing tissue below the bark and can girdle and ringbark the tree. Mostly it attacks gums, sometimes wattles. The species usually attacking *Pittosporum* may tunnel straight through smaller branches, causing them to fall.

The usual covering, resembling loosely webbed sawdust, has been removed to disclose a borer hole in Albizia. Note the bark and soft tissue eaten from around the hole.

Apart from the latter, presence can be indicated by areas of lifting bark and holes, often with gum exuding from where the adult has emerged. Repair of damage involves — lifting any loose or hollow sounding bark, using a strong water jet to clean grass from the oval tunnels, when dry puttying any deeper hole and sealing with tree paint, hoping bark will grow back over the injured area.

The tree should then be fed with a 'complete' fertiliser that has a trace element mix added, applying about one handful/m^2 of area below and a little further out than branch line. Water in well. This not only speeds recovery but it appears that the borers mainly attack ailing trees that have had the natural mulch removed from under them or trees with root problems resulting from change of soil level or diversion of water to or from the area.

An accepted theory is that normally healthy trees exude gums or kinos that engulf the borer larvae in its very early stages.

Scale Pests

These range from relatively large globular gum tree scale and white wax scale (as large as 6–7 mm across) that sometimes attack *Eriostemon*, some *Pittosporum* and *Myoporum* to the fine pinpoint size palm scale or similar type common on tea tree and some other native shrubs. Like all scale and a few other sap-sucking pests they secrete sugary substances often referred to as 'honeydew' which attract the black sooty mould fungus. This fungus lives on the secretion and is not a direct parasite of the plant. Tea tree scale can be completely hidden beneath this black mould.

The appearance of sooty mould indicates scale, although there are times when it appears on a shrub due to the sugary secretions dropping from an overhanging gum tree with scale. Generally speaking, clean up the scale and the sooty mould soon disappears.

GUM TREE SCALE

The more common one in most areas looks like cream to buff match-heads. These bulbous membranes are filled with a treacle-like fluid. This pest is often prevalent during the first few years of a tree's life after which time the tree grows out of it. Colonies can be scrubbed off with soapy water or sprayed with about 2 tablespoons of white oil plus 2 teaspoons of Lebaycid or Rogor per 4–5 litres of water, applied on a cool day.

WHITE WAX SCALE

This is more effectively treated by spraying with the white oil mixture when the young generation of scale has emerged from the wax-protected female and is mobile on the foliage as a minute buff creature usually near main veins of leaves during late November and early summer. A follow-up spraying about one month later in mid-summer is advisable. Otherwise scrub off with soapy water. The latter method is often the more prudent as various ladybirds as well as some fungi are scale predators.

Tea tree scale and other scales such as cottony cushion scale, are occasionally found on some acacias, grevilleas, etc., and can be treated as suggested for gum tree scale. However Malathion should be substituted for Rogor.

Fungus Spots and Damping Off

Fortunately these are generally less common in Australian natives than in exotics.

DAMPING OFF

This may occur in some closely matted soft stemmed plants like *Scleranthus* (Canberra grass). In this case cut out the brown or dead section, usually the centre and spray with Daconil or Benlate as directed. If the plant seems in danger of departing, cuttings can usually be taken from unaffected outer areas of the clump.

Another type of damping off fungus (pythium) sometimes attacks downy stemmed plants and particularly seedlings. Affected stems usually blacken or shrivel at soil level and the plant collapses. This usually happens when plants are overcrowded or when air circulation is restricted. The fungus can only travel through the microscopic layer of moisture over the soil surface particles, usually during darkness. Therefore, water early in the day to ensure a dry surface at night, free susceptible plants from dead underfoliage and branchlets, and make sure mulch is not too close to stems. This is one reason why a pebbly scree is recommended for such plants. Watering stems and soil as directed on the container with Fongarid should check the disease.

When sowing seed it is a good idea to use vermiculite as a surface covering. It holds water within, rather than around, the particles and its loose coarse texture foils the travel of the fungus. Also the air access it allows helps germination.

INK DISEASE OF KANGAROO PAW

Anigozanthus manglesi and some of its close hybrids are very prone to this, even in their dry summer habitats. At the first signs, remove blackened foliage and water the clump with Mancozeb, particularly during its summer dormancy period.

Black spot occurs on some *Grevillea, Hakea* and *Persoonia*. Spraying with mancozeb usually gives control although a similar spotting can be caused, especially on wattle, by the acacia bug, which is very difficult to control.

ROOT ROT

Phytophthora root rot takes its toll on a number of natives, especially those from dry summer WA areas grown on the comparatively wet east coast. The main point to remember is that this fungus can only progress in wet soil. Therefore good drainage and sensible watering play a big part in avoiding it. Fongarid, watered as directed on the container, does arrest, and for a time, prevent the disease but with most natives when initial limpness or wilting shows, it is often too late to save the plant.

It is better to have new plants coming on from cuttings and replace frequently. Practically all natives particularly prone to the disease flower quickly from cuttings or seedlings. However there are a few exceptions, including the *Verticordia* and *Eucalyptus macrocarpa.*

Propagating Native Plants

Raising at least some of your own plants from seed or cuttings can be a very satisfying and interesting experience.

Given a suitable environment and soil mix, many species reproduce easily. Others like *Persoonia pinifolia* and some of the Eriostemons, can be difficult but even with these more challenging types a little perseverance can pay dividends. Sooner or later you may strike a time when "conditions are just right".

Growing from Cuttings

The time the cuttings are taken is important. Young growth approaching maturity is usually best. However, the younger it is the more rapidly it wilts, therefore choose material that begins to lose its soft sappiness — in most species when the stem is changing from green to brown.

Type of Cuttings

Tip cuttings are the type most frequently used. These are cut 5–10 cm below the actual growing tip of the stem (not tips of top foliage). The cut is made cleanly with a sharp knife or razor blade through the base of a leaf junction. Foliage is then stripped from the lower half of the cutting. In most cases it is preferable to leave final stripping and base trimming until immediately prior to planting.

Heel Cuttings are maturing shoots from the side rather than tip of stem, usually from a leaf axil, which are gently pulled away with a slightly downward motion so that a small 'heel' of the older wood (the stem) is attached. Any appendage of bark coming away with the woody heel is trimmed back to the older wood, and the lower half of foliage is removed. Tip cuttings often form side shoots suitable for heel cuttings.

Simple materials are used to make an efficient mini-glasshouse for raising cuttings — A plastic bag; two hoops (cut from a wire coat hanger) and bent to keep plastic clear of the cuttings; a 10–12 cm plant pot and a larger one with some sand to stand it in to allow better clearance between hoops and cuttings. Alternatively, the potted cuttings may be covered by a two or three litre plastic drink bottle with the bottom cut out.

Unprepared tip cuttings of *Prostanthera* (left) and heel cuttings of *Bauera* (right)

Cuttings prepared for planting. Tip cuttings are cut cleanly through or fractionally below a leaf junction, lower foliage removed and tip leaves trimmed. Heel cuttings with bark "whisker" plus some top leaves trimmed, and lower half of foliage removed.

A number of cuttings can be struck in the one pot. A small stick a little wider than the cuttings is used to dibble a hole for each. The forked section near the top of the stick is used to gauge planting depth.

Care of Cuttings Prior to Planting

Cuttings of all but succulent-type plants should be planted if possible within 5-10 minutes from time of trimming. If this is not possible, freshly cut material should be sealed tightly in a plastic bag, with a light sprinkling of water, and kept completely shaded. This way, planting can be delayed for 2-3 days. An alternative is to wrap and cover them in several thicknesses of moistened newspaper. However, it is better to delay trimming until ready to plant, and all handling and planting should be done in a shaded and wind protected area.

Soil Mix for Cuttings

Firm contact with soil mix, moisture and oxygen are all essential to encourage root growth. Oxygen as well as water enters through the spaces between the soil particles; therefore, the coarser the particles, within reason, the greater the spaces between them. For this reason, coarse river sand is popularly used as the main component for cutting mixtures. Peat moss is added to prevent the sand from drying out too rapidly. A good standard mix is:

3 parts coarse river sand well mixed with;
1 part previously moistened peat moss.

If the river sand appears to contain powdery particles (usually silt) or appears dusty, its quality as a cutting mix will be improved by washing it, to pour or float off this finer material.

Some growers prefer to substitute leafmould, decomposed to almost soil-like consistency, for the peat moss. This does seem to contain some root stimulating properties; at least it is worth trying.

Sterilising the soil sometimes helps to offset damage from parasitic fungi, bacteria or nematodes. Partial sterilisation is preferable to complete. Partial kills off most pathogens or parasites that live on living plant tissues, but not the more useful organisms which take part in decomposition of spent plant growth or ones which inhibit the parasites. Partial sterilisation can be carried out on a small scale by filling a baking dish with cutting or seed raising soil and heating it in the oven at about 50°C (120°F) for 25-30 minutes.

Planting

Cuttings may be set in individual pots 4-5 cm in diameter, or more usually when there are a number of cuttings, set 2-3 cm apart towards the edge of a 10-12 cm pot.

Containers need to be washed clean. Fill them with the slightly damp soil mix, firm down and settle by tamping the base several times on a firm surface. If necessary add more soil mix to bring it to 1 cm of the pot rim.

Use a match-fine cane, thin pencil etc., to dibble a hole for each cutting, holding the dibbler so that the hole is no deeper than the stripped section of the cutting to be covered (say 1/3 to ½ of its length). This is important because an air space rather than firm contact with the mix at its base may result in failure.

After placing the cuttings to their full depth, firm to good contact with the soil mix by pressing from the centre of the container outwards, sandwiching them in the soil between your thumbs and the side of the container.

Water by standing the containers in about half their depth in water until there are signs of moisture creeping over the surface. However, the soil need not be completely saturated. Results are sometimes better by using a fine mist spray to bring the soil from just damp to wet, rather than saturated.

Keeping Cuttings Alive Until Roots Form

The aim is to prevent wilting by reducing natural transpiration of water from the foliage, until the cuttings make sufficient root to supply normal water requirements.

One obvious way is to remove foliage, but if completely denuded they cannot produce sugars and other carbohydrates necessary for root formation (with the possible exception of deciduous plants which store these substances in the stem before natural leaf fall). A good compromise is to trim foliage back to about half or even a third, especially the younger tip growth as this draws heavily on moisture. The best way to do this is to bunch the top foliage together and trim it back just above the growing tip. This also makes the appearance of new growth more obvious. Small foliage plants like Leptospermums, Epacris etc., do not need trimming.

Reduced light also inhibits transpiration. A bright area protected from direct sunlight is usually most suitable, although with some plants, especially shade lovers, it is better to have the area darker for the first week or so. Judge by the appearance of the cuttings. If there are signs of wilting, reduce the light.

Extra humidity is the key to success with most cuttings. The easiest way to do this is to enclose the container of cuttings in a plastic bag. To keep the bag clear of cuttings, it is a good idea to place the container in another a few sizes larger, with clean sand in the larger pot to keep the cutting container in place and bring its level to that of the larger one. For overhead support, place four light sticks in the sand around the container, high enough to keep the plastic free of the cuttings, or better still, cross two wire hoops above the cuttings. Wire coathangers are ideal, as they bend easily to spring fairly firmly against the side of the outer container. Large plastic drink bottles with their base removed also make excellent miniature glasshouses.

With cuttings firmed and watered gently, the container is supported to the right height with damp sand in the larger pot. Hoops are positioned and covered with the plastic bag and secured under the larger container or held with a large elastic band.

Cuttings will form roots from about a few weeks to a few months depending on species, the stage of maturity of cuttings taken, temperature, and conditions generally. Partly remove the plastic cover after a few weeks. Watch for signs of wilting, and if it appears, replace the cover. Only after removal of the cover should the cutting be gradually conditioned to full sunlight. Remember that this glasshouse-type condition is intended to create only humidity, not extra heat.

Many downy foliaged or fleshy leafed plants, like Geraldton Wax, will rot if humidity is excessive. These need only shade and shelter from wind. If in doubt, try some under both conditions.

Check for root formation by gently removing one or two of the cuttings, or, if uncertain, carefully tap the slightly damp soil ball from the container to examine a few. Repot into individual containers before roots become entangled. Repot in shade, keep the plants shaded for a few days, then gradually condition to normal sunlight.

The bottomless bottle is an easy alternative. This may be opaque or clear plastic.

When repotting, your need is to grow plants that will adapt to your soil conditions. Therefore, aim for the happy medium, a soil close to that of your garden but crumbly enough to encourage good root growth and drainage, holding enough moisture to need no more than daily watering.

A good mixture would be:
7 parts by volume of your garden soil;
2 parts coarse river sand;
1 of moistened peat moss and;
1 of well-decomposed compost or leafmould.

If your soil is inclined towards a heavy loam, then add another two parts of sand. If sandy, one extra part of peat moss, as you would need to improve the soil of the planting area.

Container size should be increased gradually, starting with 5 cm or no more than 10 cm pots. When roots begin to show around the soil ball, transfer to only about two sizes larger or better still, plant into permanent positions while the plants are small.

Seed Sowing

There are a number of good proprietary seed raising mixtures available or, alternatively, use the cutting mix suggested, plus an extra part of peat moss to prevent quick drying. The use of a plastic cover may be needed to prevent drying until germination takes place. Once seedlings emerge, the cover should be removed and normal sunlight introduced to prevent growth becoming weak and spindly.

Sow seed thinly in well-spaced rows to ensure stronger and healthier seedlings (crowding results in weak, leggy growth). Cover rows with white sand or vermiculite which acts as a marker for the sown areas, and makes it easier to distinguish seedlings from weeds. Vermiculite also holds moisture without retarding entry of air, and minimises damage from damping off fungi if used as a complete surface cover.

Protection from heavy rain or watering and some break from hot sunlight can be made by folding a sheet of fibreglass insect gauze into a tent-like formation to enclose the seedling container. Clothes-pegs can be used to hold the folds. It also provides protection from slugs, snails and other chewing insects.

Sow the seed in wide but not necessarily deep containers so that the seed are well spaced, preferably in rows at least 1 cm apart. Thin out or space the tiny seedlings as soon as they are large enough to handle. Leaving them close together encourages damping off fungi as well as legginess. Cover seed to about twice its depth or if using vermiculite, up to 1 cm deep.

Be patient. Some seed germinates within days, other types months, and some spasmodically. Space them out or pot into individual containers as soon as possible.

An easily made "tent" from insect screening will protect newly raised seedlings from damage by heavy rain or watering, wind, slugs and snails, chewing insects, and birds.

Acacia

In the vicinity of 700 species are native to Australia, varying in form from small shrubs to medium size trees. From the time of Australia's first settlement, acacias have been known as wattles because their flowers resemble those of Callicoma (Creek Wattle) so called because it grew along the Tank Stream which flowed into Sydney Cove. Having long, pliable but strong, stems the latter plant was used for weaving walls of the early 'wattle and daub' huts.

All acacias display generous clusters of tiny florets, each carrying numerous filaments or stamens separated or 'free' to their base. The florets are united into fluffy balls or fingers on single stemlets (peduncles). Colour is more commonly golden yellow, lemon or occasionally creamy white.

All form strap-like fleshy seed pods, often constricted between seeds. Leaves are alternate, sometimes bipinnate and fernlike or modified into oval or strap-like phyllodes. Seedlings often start with finely divided leaves then change to comparatively large leaf-like phyllodes typical of the adult species.

Acacia podalyriifolia

Some wattles are long lived but others may have an average life of only five to ten years but these are rapid growers, therefore they can be quickly replaced. Also the shorter lived species are useful to fill in as screens or shelters while slower growing trees are establishing. Apart from the brilliance of their flowers, the silver foliaged species like *Acacia baileyana, A. binerva (glaucescens), A. cyanophylla, A. podalyriifolia* and others provide attractive foils for all other colours and give an illusion of depth to a landscape or distance when used towards the end of vistas.

Acacia fimbriata

Acacia accola

See *A. adunca.*

Acacia adunca

(syn. *A. accola*)
Wallangarra Wattle
FAMILY MIMOSACEAE

Large shrub or small round headed tree carrying slightly drooping slender foliage. Spectacular with its canopy of bright golden flowers during late winter or early spring.

Origin East coast Australia, mainly northern Vic. and southern NSW.
Size 4–7 m high, top spread of 3–4 m.
Soil Any average well-drained soil.
Climate Cooler temperate to tropical; frost tolerant.
Aspect Needs at least half sun for good flower display.
Water Moderate. Good drought resistance once established.
Feeding Light dressing of blood and bone in spring will speed growth.
Pruning Tip prune during younger stages if bushier growth is preferred.
Propagation Scarified or hot-water-soaked seed.
Possible Problems Rarely occur.

Acacia accola

Acacia amblygona

Acacia amblygona

FAMILY MIMOSACEAE

A semi-prostrate tangle of blue-green stems with small foliage 1–2 cm long, varying from lance to oval shape and a liberal scattering of bright yellow ball flowers from mid winter into spring. A good groundcover, especially a comparatively dwarf form from southern Qld which rarely exceeds 30 cm in height.

Origin Qld down to central NSW coast.
Size To 1.5 cm high (also see dwarf form above) width 1–1.5 m.
Soil Any well-drained soil, with preference for a well mulched sandy loam.
Climate Semi-tropical to temperate with some protection from severe frost.
Aspect Full sun to dappled sunlight.
Water Moderate, at least until established.
Feeding A little blood and bone acceptable in spring but avoid high nitrogen fertiliser.
Pruning May be lightly pruned as flowers finish and any taller erect shoots pruned back.

Propagation Scarified seed.
Possible Problems Browning of underfoliage in soils that remain moist.

Acacia aneura

Mulga

FAMILY MIMOSACEAE

A large shrub or small tree with rough grey-brown flaky bark. Slender grey phyllodes to 25 cm long. They can vary from flat, linear, 5 mm wide to needle-like forms, usually upright but in some areas pendulous. The bright golden yellow rods of flower are 3–5 cm long, mainly from mid-winter to mid-spring. A good light shadetree. Stock will eat foliage. The close grained wood polishes well — popular for souvenirs.

Origin Inland areas of all mainland States.
Size 4–10 m high 2–6 m wide.
Soil Most well-drained soil, but preferably crumbly or light loam. Accepts alkaline conditions.
Climate Temperate to semi-tropical — best suited for semi-arid conditions. Good frost tolerance.
Aspect Preferably open — but accepts up to half shade.
Water Moderate until established. Good drought tolerance.
Feeding Low fertility tree but occasional slow release, complete or animal manure speeds growth.
Pruning Rarely necessary but is acceptable.
Propagation Seed.
Possible Problems Animals seem partial to foliage but otherwise fairly trouble free.

Acacia aneura

Acacia baileyana

Cootamundra Wattle

FAMILY MIMOSACEAE

Finely divided silver-grey fern-like foliage almost eclipsed during late winter–early spring with pendulous sprays of small golden yellow balls of flower. Average life span five to ten years but growth is rapid, making it an excellent temporary screen while slower growth establishes.

Origin Probably northern Vic. and southern NSW but now widespread.
Size Height 5–7 m, width 4–5 m.
Soil Any well-drained soil.
Climate Cool to semi-tropical.
Aspect At least half sun.
Water Moderate until established.
Feeding Speed establishment with a light dressing of blood and bone in spring.
Pruning Improved by light pruning immediately after flowering; if possible avoiding cutting into older branches.
Propagation Hot-water-soaked or scarified seed.
Possible Problems Borer, especially in warm to temperate areas.

Acacia baileyana

Acacia boormanii

(syn *A. hunterana*)

Snowy River Wattle

FAMILY MIMOSACEAE

Shrub with slender strap-like greyish green foliage to 10 cm long, and dense clusters of fluffy golden balls in spring.

Origin Southern NSW and Vic.

Size Height 3–4 m sometimes taller in shelter, spreading 2–3 m.
Soil Preference for medium loam but adaptable if drainage is good.
Climate Cold to semi-tropical.
Aspect Full sun to half shade.
Water Good drought resistance once established.
Feeding Optional feeding with blood and bone or native plant food in spring.
Pruning Tip prune young plants for lower bushier growth — prune lightly after flowering.
Propagation Scarified or hot-water-soaked seed. Tip or heel cuttings in late summer may be successful.
Possible Problems Rare.

Acacia chinchillensis

Acacia chinchillensis

Darling Downs Wattle

FAMILY MIMOSACEAE

A decorative and adaptable small spreading shrub with dark glossy green bipinnate leaves, the oblong pinnules (leaflets) less than 1 cm long, giving a soft fern-like appearance. Their dark colour contrasts with the fluffy globular bright golden yellow flower heads which appear in late winter to early spring.

Origin Southern Qld, especially Chinchilla area.
Size 0.7–2 m high and a similar width.
Soil Adapts to most well drained light to medium loams.
Climate Cooler temperate to tropical.
Aspect Dappled to full sunlight.
Water Moderate, at least until established.
Feeding Light dressing with native plant food, complete or slow release fertiliser acceptable in spring.
Pruning Light pruning after flowering encourages bushier and stronger growth.
Propagation Seed.
Possible Problems Usually trouble free.

Acacia boormanii

Acacia cultriformis

Dog Tooth Wattle

FAMILY MIMOSACEAE

A fairly erect shrub with bluish green nearly triangular foliage crowding stems and branches. Carries a vivid canopy of deep golden ball flowers in spring.

Origin Mainly west of Dividing Range in southern Qld and NSW.
Size 2–4 m high — 2–3 m wide. There is also a prostrate or gracefully trailing form.
Soil Prefers gritty or gravelly well-drained soil.
Climate Cold to semi-tropical. Flowers best in areas with cold nights. Suits inland districts. Frost tolerant
Aspect Best in open sunny situations.
Water Moderately until established then has good drought resistance.
Feeding Optional light dressing of blood and bone or animal manure in spring.
Pruning Prune off flower heads as they finish.
Propagation Hot-water-treated or scarified seed.
Possible Problems Usually trouble free, especially in dry climates.

Acacia decurrens

Black Wattle

FAMILY MIMOSACEAE

Dark trunked high branching tree with rounded crown of dark green finely

Acacia cultriformis

Acacia cultriformis (prostrate form)

divided feathery foliage almost covered with cloud-like masses of small bright yellow ball-shaped flower clusters in late winter–spring. In some districts borers reduce life to seven or eight years, but plants re-establish quickly. A copse of three or five trees makes an appealing natural shade area.

Origin NSW and Qld but naturalised throughout eastern Australia.
Size 6–10 m high with an above head height spread of 4–6 m.
Soil A preference for crumbly loams or light shale soils but also adapts to sand — needs reasonable drainage.
Climate Cold to tropical.
Aspect Best in fairly open sunny situations.
Water Prefers moderate moisture but good drought resistance when established.
Feeding Light dressing blood and bone or native plant food optional in spring.
Pruning Not practical.
Propagation Scarified seed.
Possible Problems See description above.

Acacia drummondii

FAMILY MIMOSACEAE

A comparatively low growing shrub with blue-green fern-like foliage and 2–3.5 cm long broad fingers of soft golden flowers during late winter to mid spring. There are several varieties including var *major* which has the largest flowers and var *affinis*, which usually has hairy leaflets as opposed to smooth foliage.

Origin Southern WA.
Size Usually less than 1 m but occasionally to 2 m high and to 1.5 m wide.
Soil Prefers well drained gravelly our gritty soil with compost added. Crushed shale or sandstone as a base. Mulch should help in heavy loams.
Climate Temperate to cooler temperate but resents severe frost.
Aspect Dappled sunlight or some sun protection during the hottest part of the day.
Water Moderate, at least until well established.

Acacia drummondii

Acacia decurrens

Feeding Very light dressing of blood and bone in spring.
Pruning Prune back lightly after flowering.
Propagation Scarified seed or heel cuttings in late summer.
Possible Problems Fungus may attack lower foliage if in a sheltered humid area.

Acacia elata

Cedar Wattle

FAMILY MIMOSACEAE

One of the fastest growing but fairly long lived acacias with erect trunk clad in lateral side branches and a dense cover of deep green frond-like leaves divided into eight to twelve pairs of slender elliptical leaflets. Clusters of medium to large cream flower balls in early summer.

Origin Mainly coastal and mountain areas of Vic. and NSW.
Size 10–25 m high 7–2 m wide
Soil Any well-drained soil.
Climate Cool to tropical districts, least suited to dry inland.
Aspect Grows well in shaded gullies or exposed areas.
Water Needs at least moderate water for rapid growth.
Feeding Blood and bone or soluble fertiliser in moderation during spring and early summer will speed growth.
Pruning Old flower heads may be trimmed off where accessible.
Propagation from scarified seed.
Possible Problems Usually trouble free.

Acacia fimbriata

Brisbane Wattle or Fringed Wattle

FAMILY MIMOSACEAE

SEE PAGE 33

An attractive small tree with rounded bushy growth. The slender dark green phyllodes are 2–5 cm long, fringed with minute hairs on the edges (fimbriate). From late winter to early spring the tree is massed with delicately fragrant bright yellow balls of flower.

Origin Qld and NSW.
Size 5–7 m high and nearly as wide.
Soil Adapts to a wide range of moderately moist but well-drained soils, preferably with a generous surface mulch.

Acacia elata

Climate All but the coldest mountain climates
Aspect From full sun to light shade, best with some shelter from severe winds.
Water Moderate, especially in summer — resents prolonged drought.
Feeding Needs little but surface mulch. Slow release or complete fertiliser applied moderately in spring will speed growth.
Pruning If practicable,benefits from light pruning after flowering.
Propagation From scarified seed
Possible Problems Not usually affected although borer damage can occur.

Acacia glaucescens

Coastal Myall

FAMILY MIMOSACEAE

Small to medium round topped tree with rough brown bark and a dense crown of silver-grey broad lance-shaped phyllodes to 15 cm long, making delightful foil for the rods of golden spring flowers. Leaves reputedly poisonous to stock.

Origin Mainly from NSW coastal areas.
Size 12–15 m high 5–7 m wide (above head height).
Soil Prefers lighter to sandy loam but adapts where drainage is good.
Climate Temperate to tropical areas.
Aspect At least half sun and not too crowded.
Water Moderate watering, at least until established.
Feeding: Light dressing of blood and bone or fowl manure in spring (optional).
Pruning Not practical but tip pruning encourages a bushier and lower branching plant, at least for the first five years or so.
Propagation Scarified seed.
Possible Problems Usually trouble free.

Acacia glaucoptera

Flat or Saw-tooth Wattle

FAMILY MIMOSACEAE

An unusual wattle with alternate blue-green tooth shaped wings flattened along a midrib to create a double edged saw formation. New growth is often bronze-red. Flowers are bright yellow balls to almost 1 cm. in diameter on short reddish stems from nodes along the midrib during late winter and early spring.

Origin Southern WA, particularly from Stirling Ranges to Esperance
Size 0.5–1.5 m high, to 2 m wide.
Soil Adaptable, as it grows naturally both on clayey flats, gravel and crumbly red soil of Mallee areas — needs moderate drainage.
Climate Cooler temperate to semi-tropical. Some frost tolerance.
Aspect At least half sun — suits lightly broken sunlight.
Water Moderate — good drought resistance when established.
Feeding Lightly with blood and bone or complete plant food in spring.
Pruning Prune back by about a third after flowering.

Acacia glaucescens

Propagation Scarified seed or from cuttings.
Possible Problems Rarely affected.

Acacia longifolia

Sally Wattle
FAMILY MIMOSACEAE

Large shrub to small tree with smooth light green phyllodes to 15 cm long and sometimes to 3 cm wide. These are intermingled with a profusion of bright yellow flower rods during spring. Fast growing and long lived.

Origin Eastern Australia.
Size 6–8 m high 4–6 m wide. Var. *sophorae* (with shorter phyllodes) grows only 1–3 m high but to 4 m wide.
Soil Adapts to most well drained soil but excellent for reclaiming sand dune areas.
Climate Cool frosty to tropical areas. Excellent salt wind resistance.
Aspect Accepts light shade but outstanding in exposed windy situations.
Water Moderately until established.
Feeding Accepts poor sandy soil but a light dressing of blood and bone or fowl manure in spring speeds growth.
Pruning Outer branchlets may be pruned back to about a third length as flowers finish.
Propagation Scarified seed.
Possible Problems Good resistance to pests and diseases.

Acacia multispicata

FAMILY MYRTACEAE

A very showy wattle with stiff grey to green slender pine like phyllodes varying from 4–8 cm long and numerous fluffy fingers of bright yellow flower heads to 3 cm long from mid-winter to mid-spring.

Origin Southern WA.
Size 2–3 m high and nearly as wide.
Soil Prefers gritty sand or gravelly soil but adapts to composted well-drained loam.
Climate Cooler temperate to semi-tropical but resents severe frosts, especially until established.
Aspect Fairly open position with at least half sun.
Water Moderate — allow surface to dry between waterings.
Feeding Light dressing of blood & bone or fowl manure in spring.
Pruning Prune back by about one-third after flowering.
Propagation Scarified seed.
Possible Problems Usually trouble free.

Acacia murrayana

Murray's Wattle
FAMILY MIMOSACEAE

An ornamental medium to large shrub

Acacia glaucoptera

Acacia longifolia

Acacia multispicata

Acacia murrayana

with soft pink to white sprawling trunk which then breaks into numerous upright silvery mauve stems fairly densely clad in club- to lance-shaped but round-pointed grey-green phyllodes from 6–15 cm long. Bright yellow to deep golden balls of flower heads on long peduncles 5–7 mm across, in short racemes from late winter to mid-spring.

Origin Inland areas of all mainland States.
Size 2–6 m high,usually 2–4 but occasionally up to 6 m wide.
Soil Preferably sandy to gravelly but adapts to most well-drained alkaline or mildly acid loams.
Climate Ideal for hot dry semi-arid to arid inland areas — stands heavy frost.
Aspect Best in open sunny situation — good wind resistance.
Water Moderate until established then drought resistant.
Feeding Light sprinkling of complete plant food or scattering of fowl manure is acceptable after 2–3 months from planting.
Pruning May be pruned back to about half after flowering to thicken growth.
Propagation Seed.
Possible Problems Usually trouble free but may be short lived in frequently moist soils.

Acacia oxycedrus

FAMILY MIMOSACEAE

An erect, prickly, usually bushy shrub with stiff, sharply pointed grey-green dagger-shaped phyllodes from 2–4 cm long and soft creamy yellow slender cylindrical flower spikes 2–4 cm long from early winter to spring.

Origin Coastal NSW and Vic. into SA.
Size 2–3 m high and nearly as wide but reported to reach small tree proportions in some areas.
Soil Has preference for sandstone ridges but reasonably adaptable.
Climate Temperate to cool temperate — good frost resistance.
Aspect Needs at least half or only lightly broken sunlight.
Water Moderate until established — good drought resistance.
Feeding A leafy mulch seems sufficient.
Pruning May be pruned back after flowering without denuding branches of foliage.
Propagation From scarified seed.
Possible Problems Usually trouble free.

Acacia podalyriifolia

Queensland Wattle

FAMILY MIMOSACEAE
SEE PAGE 33

Large upright but bushy low-branching shrub or small tree with downy silver-grey oval foliage and clusters of large golden ball flowers during winter.

Origin Queensland but naturalised in most States.
Size 4–6 m high and 4–5 m wide.
Soil Any well drained soil.
Climate Cool to tropical.
Aspect At least two-thirds sun.
Water Moderate, at least until established.

Acacia oxycedrus

Acacia rigens

Acacia pycnantha

Feeding Optional light feeding with blood and bone in spring.
Pruning Flower carrying stems or branchlets may be reduced by about one-third after flowering.
Propagation From scarified seed.
Possible Problems Average life 7–10 years but rapid grower and good temporary screen. Unfortunately leaf miners mar foliage in some coastal districts.

Acacia pycnantha

Australian Golden Wattle
FAMILY MIMOSACEAE

Australia's floral emblem — large shrub or small tree with bold shining green foliage, sickle shaped to 15 cm long, widening to 5 cm above the centre and tapering to a rounded tip. Leaves are often largest when young. The fragrant large golden balls of flower to 10 mm diameter come in spectacular masses during spring.

Origin SA, Vic. and southern NSW but widely naturalised.
Size 4–10 m tall 2-3 m wide.
Soil Well drained loam but adapts to sand.
Climate Semi-tropical to cooler temperate areas. Good frost tolerance.
Aspect Full sun or lightly dappled shade.
Water Appreciates ample water but reasonable drought resistance.
Feeding Light dressings of blood and bone in spring will speed growth. Appreciates a leafy mulched surface.
Pruning Prune to remove old flower heads if practicable.
Propagation Scarified seed.
Possible Problems Usually trouble free.

Acacia rigens

Needle Bush Wattle
FAMILY MIMOSACEAE

A fairly open shrub with greyish green stiff needle-like foliage to 16 cm long, ascending rather than pendulous and tapering to a sharp point. During late winter or spring it is brightly decked with large golden balls of flower on peduncles from the leaf axils.

Origin Eastern and southern Australia.
Size 2-4 m high and as wide.
Soil All but the heaviest well-drained loamy soil.
Climate Seems to suit all but tropical monsoonal areas. Good drought and frost tolerance.
Aspect Full to part or broken sunlight. Tolerance to strong winds.
Water Moderate, at least until established
Feeding Slow release or complete fertiliser applied moderately when soil is moist in late spring will speed growth.
Pruning Not necessary but may be pruned back by about one-third after flowering.
Propagation Scarified seed
Possible Problems Rarely affected.

Acacia saligna

(syn. *A. cyanophylla*)
Golden Wreath Wattle
FAMILY MIMOSACEAE

Acacia saligna

A fast growing small tree, often with spreading umbrella-like or gracefully

Acacia spectabilis

Actinodium cunninghamii

pendulous crown. Long slender phyllodes to 22 cm long, variable width from 1-2 cm, dark green or occasionally blue-green — brilliant in late winter to spring with large golden balls of flower clustered along the pendulous outer branches. One of the more spectacular and most vigorous wattles with a reasonable life span — so vigorous that it becomes invasive in some areas,especially along the eastern coast of Australia. However, it is a good small shade tree.

Origin WA but naturalised in all States.
Size 4-8 m high with a similar crown spread.
Soil Adapts to any moderately drained soil.
Climate Cool to tropical. Good frost and salt wind tolerance.
Aspect Full to half sunlight.
Water Accepts liberal watering but good drought resistance when established.
Feeding Light sprinkling of complete plant food or fowl manure in spring speeds growth but accepts low fertility soil.
Pruning May be pruned back to about half after flowering if needed for shaping.
Propagation Scarified seed.
Possible Problems Mildly subject to borer but usually trouble free.

Acacia spectabilis

Mudgee Wattle

FAMILY MIMOSACEAE

Small tree with slender erect trunk. The slightly pendulous outer branchlets display attractive fern-like blue-green foliage divided into tiny oval leaflets which contrast beautifully with the generous clusters of golden ball flowers in spring.

Origin NSW and Qld, mostly west of the Great Dividing Range.
Size 5-6 m high, lower branches to 3 m usually tapering to slender apex.
Soil Adapts to most well-drained soil.
Climate Temperate to semi-tropical, good frost resistance.
Aspect At least half sun, preferably with some protection from strong winds.
Water Good drought resistance once established.
Feeding Light dressing of blood and bone acceptable in spring.
Pruning Outer branches may be pruned back lightly to remove spent flowers in spring.
Propagation Scarified seed.
Possible Problems Usually trouble free.

Actinodium cunninghamii

Swamp Daisy

FAMILY MYRTACEAE

A small twiggy plant with scale-like, overlapping, stem-clasping leaves to 5 mm long, and carrying showy terminal heads 4-5 cm across, of florets arranged in the formation of a double daisy, the centre florets rosy red, the outer and larger sterile ones white. These appear mainly from late winter to mid-summer.

Origin Southern WA, especially on lowlands surrounding the Stirling Range.
Size 30-35 cm high and spreading 30-40 cm.
Soil Irrespective of the common name, they also accept well-drained but reasonably moist sandy to medium loam.

Climate Temperate. Tolerant to light frost.
Aspect Preference for lightly broken sunlight but accept full sun.
Water Moderate until established then occasional soakings during prolonged dry periods.
Feeding Light scattering of slow release or complete fertiliser.
Pruning Trim back to about two-thirds after flowering.
Propagation Cuttings.
Possible Problems Usually trouble free.

Actinotus helianthi

Flannel Flower
FAMILY APIACEAE

A soft wooded downy plant with deeply cut silver-grey foliage hugging the erect white stems. These support slender branching sprays of creamy white daisy-like flowers to 8 cm across with cleanly cut flannel-textured bracts surrounding a woolly centre.

Origin NSW especially sandy areas of the central and south coast.
Size 50–80 cm high to 50 cm wide.
Soil At home in sand enriched with rotted leafmould but adapts to most well-drained light to medium loams.
Climate Temperate.
Aspect Dappled sunlight to full sun. If in sandy soil and full sun, protection by mulch or lower growing plants is needed.
Water Moderate — surface needs to dry out between waterings.
Feeding Blood and bone or slow release plant food applied moderately in spring and autumn.
Pruning Plants may be cut back to about one-third after flowering.
Propagation Seed.
Possible Problems Red spider mites can cause yellowish mottling of foliage or frequent wet soil surfaces induce fungus rots. A gritty surface scree minimises the latter.

Actinotus minor

Miniature Flannel Flower
FAMILY APIACEAE

An insignificant but to some extent cute little plant with tiny foliage and long thread-like stems supporting minutely bracted flowers no more than 1 cm across. Flowers appear mainly during spring and summer. Adds interest as an undercover plant, accompanying tetratheca or other low growing plants.

Origin Mostly coastal sandstone areas of NSW.
Size To 20 cm high, 25–30 cm wide.
Soil Most well-drained soils with leafy or rocky mulches.
Climate Temperate.
Aspect Prefers lightly broken sunlight but accepts full sun.
Water Moderate.
Feeding Leafy mulch is sufficient.
Pruning Old flowers may be trimmed off.
Propagation Seed.
Problems Rarely affected.

Actinotus minor

Actinotus helianthi

Agonis flexuosa

Willow Myrtle, Peppermint

FAMILY MYRTACEAE

A graceful, usually flat topped tree with deep green slender foliage to 10 cm long hanging from gracefully arching outer branches. The exposed upper sections of these branches are studded with small white tea-tree-type flowers during late spring or early summer. Juvenile foliage is broader.

Origin WA.
Size Usually about 5 m high and 4–5 m wide and branching above head height. Some old established specimens to 12 m.
Soil Adapts to most well-drained soils.
Climate Semi-tropical to cool but resents severe cold.
Aspect Full to half sun.
Water Moderate until established then good drought resistance.
Feeding Soluble nitrogenous or slow release fertiliser during spring and summer will speed growth.
Pruning In winter, but only if necessary.
Propagation Seed or cuttings
Other Varieties A dwarf foE growing only about 1 m and as wide with bronze to green broad juvenile foliage is available. Also 'Variegata,' a gracefully pendulous cream variegated form which grows well in Melbourne gardens but difficult to propagate from cuttings.
Possible Problems Generally trouble-free but leaf webbing caterpillars are a possibility.

Agonis flexuosa

Albizia lophantha

Cape Wattle

FAMILY MIMOSACEAE

A quick growing small tree with finely divided fern-like foliage and greenish cream feathery bottle-brush-type flowers from winter to late spring.

Origin Southern WA, often near watercourses.
Size About 5 m high, usually only 2-3 m wide.
Soil Prefers medium to heavy loam but also grows well in acid sandstone east coast regions.
Climate Cooler temperate to semi-tropical.
Aspect Full to dappled sunlight.
Water Grows in relatively dry areas but growth more rapid with fairly frequent watering.
Feeding Light dressing of blood and bone in spring.
Propagating Seed — usually flowers within two years.
Possible Problems Usually trouble free.

Alyogne heugelii

(syn. *Hibiscus heugelii*)

FAMILY MALVACEAE

A fairly open, rounded shrub with leaves divided usually into three lobes, sometimes with each lobe almost entire or with secondary lobes. The hibiscus-like flowers about 12 cm across are usually mauve with rosy purple markings or sometimes soft lavender blue, occurring during spring and summer.

Origin Southern and western Australia.
Size About 2 m high and 1.5 m wide.
Soil Adapt to most well drained soils.
Climate Cooler temperate to semi-tropical — moderate frost resistance.
Aspect At least half but preferably full sun.
Water Moderate — good drought resistance when established.
Feeding Complete plant food, fowl manure or blood and bone in early spring.
Pruning Shorten back stems when flowers finish or leave heavier pruning until winter.
Propagation From tip or heel cuttings or seed.
Possible Problems Not usually affected but plants become woody or leggy and are best replaced after about five years.

Angophora hispida

(syn. *A. cordifolia*)

Dwarf Apple

FAMILY MYRTACEAE

A base-suckering clump of branching stems with stiff oval to oblong leaves, 13–12 cm long, new growth and flower buds covered with silky crimson hairs; Broad clusters of cream flowers about 4

Albizia lophantha

Alyogne heugelii

cm. across with sweet apple aroma during early summer, followed by large seed capsules. When in flower it is a wonderful source for beetle collectors.

Origin NSW mainly higher coastal and mountain sandstone ridges.
Size 3–5 m high, 3–4 m wide.
Climate Cool temperate to semi-tropical.
Aspect Open sunny to lightly broken sunlight.
Water Moderate.
Feeding Natural leafy mulch usually sufficient or light scattering of fowl manure or blood and bone in early summer.
Pruning Old flower heads may be pruned off if desired.
Propagation Seed.
Possible Problems Usually trouble free.

Angophora hispida

Anigozanthus

Among the most interesting of Australia's unique flora, initially classified under the Amaryllidaceae family but now more specifically as Haemodoraceae. They make rhizomaceous clumps of strap-like glossy foliage from only 12-15 cm high to 1 m or more according to species and a similar width. Height of flower spikes similarly varies from about 30 cm in such as *A. humilus* to 3 m high branching stems in *A. flavidus*.

Some of the recently developed hybrid anigozanthus are larger still, fifteen or more branching spikes per clump towering over 3 m high. However it is for vigour and adaptability that these new hybrids are so outstanding. Some in fact are more compact than the species but have the great vigour and wet soil tolerance of *A. flavidus* combined with the brighter colours of *A. bicolor*.

They also produce more flower spikes and over a longer period with much greater resistance to 'ink' disease which weakens and shortens the life of WA's striking red and green floral emblem, *A. manglesii*. Raising these hybrids from tissue culture also eliminates the debilitating diseases that could otherwise be passed on by vegetative propagation.

Anigozanthus bicolor

Anigozanthos bicolor

FAMILY HAEMODORACEA

Very similar to *A. manglesii*, WA's well-known red and green floral emblem but with the fused petals, or 'paw, being a little more golden and smaller. Flowers mainly in September.

Origin From Albany to north of Perth, WA.
Size Stems to about 50 cm high, clump almost as wide.
Soil Grows on both sand and clay, seems happiest in well-composted sandy loam with good drainage.
Climate Temperate to cool temperate.
Aspect Lightly broken sunlight or morning sun.
Water Moderate — accustomed to wet winters.
Feeding Native plant food or such as rose food applied in autumn.
Pruning Cut off old flower spikes and dead foliage.
Propagation Seed or division of clumps in autumn.
Possible Problems Possibility of 'ink' disease; see *A. manglesii*.

Anigozanthus flavidus

Albany Kangaroo Paw

FAMILY HAEMODORACEAE

The most vigorous, adaptable and disease resistant species. It has donated vigour to the new hybrid strains. It makes dense 1 m high clumps of 3-4 cm wide bright glossy green leaves with branching flower stems 2.5 m tall, carrying masses of hairy flowers. More usually yellowish green but also ranging through orange-buff to deep red. Flowers dry well for long lasting indoor decoration.

Origin Southern WA.
Size See description.
Soil Adapts to most soils from fairly dry sand to boggy badly drained clay.
Climate Cooler temperate to semi-tropical.
Aspect At least half sun or lightly broken sunlight.
Water Prefers ample water but adapts to moderately dry conditions.
Feeding Poultry manure, complete plant food or slow release fertilisers during autumn.

Anigozanthus flavidus

Anigozanthus humilis

Pruning Remove old flower stems or cut for indoors when mature.
Propagation Seed, flowering in third season, or divide clumps in autumn.
Possible Problems Usually trouble free, except perhaps for slugs and snails but can develop ink disease — see *A. manglesii.*

Anigozanthus humilis

Cat's Paw
FAMILY HAEMODORACEAE

A deciduous perennial with clumps of 12–15 cm high tapering sickle shaped foliage and spikes to 45 cm high, crested with dense heads of downy florets, each 4–5 cm long, resembling cat's paws with claws extended. These occur mainly from early winter to late spring, usually with reddish and creamy yellow tones but creamy white, yellow and deep russet tonings also occur.

Origin Temperate coast and west almost to desert region of WA.
Size As above.
Soil Prefers well-drained composted sand, also adapts to limestone and medium loams.
Climate Cooler temperate to semi-tropical.
Aspect Lightly broken sunlight or full sun with light protection during hottest part of day.
Water Moderate watering, particularly in autumn and winter.
Feeding Native plant food, blood and bone or light dressings of fowl manure in autumn.
Pruning Trim off old flower spikes.
Propagation Division of clumps in autumn or can flower in about eighteen months from seed.
Possible Problems See *A. manglesii.*

Anigozanthus manglesii

Red and Green Kangaroo Paw
FAMILY HAEMODORACEAE

WA's spectacular floral emblem.

Clumps of bluish green 2–3 cm wide foliage to about 30 cm high. Woolly purplish flower stems from 1–1.5 m changing abruptly to bright red just below lower florets and extending through to the flower calyx, then changing again to a one-sided broad bright green inflorescence, appearing mainly from early to late spring. Life span seems only four to five years or less in humid districts.

Origin Southern WA and north to Geraldton area.
Size See above.
Soil Well-drained composted sand or gravelly soils.

Anigozanthus manglesii

Anigozanthus 'Regal Claw'

Anigozanthus viridis

Climate Cooler temperate to semi-tropical.
Aspects Very lightly broken sunlight or morning sun.
Water Mainly during winter. Frequent summer watering seems to accelerate ink disease problem, although plants become deciduous under dry summer conditions.
Feeding Blood and bone, native plant food or animal manure during late autumn.
Pruning Remove old flower heads and blackened or dead leaves.
Propagation Flowers from seed in about 18 months (late summer to autumn sown). Also by division of clumps in late autumn although plants from latter may be more disease prone.
Possible Problems Plants become deciduous naturally without summer watering. They seem to have a natural life span of four to five years but this is shortened by ink disease and damping off in moist summer districts. Ink disease is most severe in this species but some control is possible by spraying Zineb or Mancozeb at first signs of blue-black foliage discolouration. Hybrid types usually have a little more resistance and vigour.

Anigozanthus viridis

Green Kangaroo Paw
FAMILY HAEMODORACEAE

One of the smaller growing kangaroo paws with leaf clumps about 25 cm high, and 50 cm stems carrying a crest of 5 cm long downy bright metallic green or sometimes golden flowers from early to late spring. Semi-deciduous in summer.

Origin From north of Perth to south of Bunbury, WA, usually in moist or swampy soils.
Size See description.
Soil At home in moist peaty soil but adapts to average moderately moist well drained light loam.
Climate Temperate:
Aspect Lightly dappled sunlight or about half sun.
Water At least moderate.
Feeding Light dressing of fowl manure or complete plant food during autumn.
Pruning Remove spent flower heads.
Propagation Flowers from seed in about eighteen months or divide crowns in autumn at first signs of new growth.
Possible Problems Ink disease or damping off may cause losses during summer dormancy, especially in districts with moist humid summers.
See *A. manglesii.*

Anigozanthus 'Dwarf Delight'

Anthocercis littorea

FAMILY SOLANACEAE

A small shrub with fleshy and sometimes sticky club-shaped (obovate) leaves 2–5 cm long. The unusual flowers 5–7 cm across have five spidery creamy yellow petals joining into a slightly funnelform purple striped throat. These appear in spring and summer.

Origin WA, mainly to the north and west of Perth, also south along the Darling Range area.
Soil Limestone, limey sand or light loam — well drained.
Climate Temperate, preferring dry summer atmosphere — reasonable frost resistance.
Aspect Full sun to lightly dappled sunlight.
Water Moderately — resents frequently wet soil during summer.
Feeding Light dressing of blood and bone or animal manure during autumn.
Pruning Tip prune when young then lightly after flowering.
Propagation Cuttings.
Possible Problems Rare — growth sparse and spindly if too shaded and poor in acid soils.

Astartea fascicularis

FAMILY MYRTACEAE

A small shrub with fine deep green heath like foliage, generously spangled with dainty five-petalled pink to white flowers to about 12 mm in diameter carried from mid-winter to late summer.

Origin Southern WA.
Size 1 m high, 1–1.5 m wide. There are also prostrate forms.
Soil At home in moist peaty sand but adapts to all but heavy clay unless lightened with leafmould etc.
Climate Cool to semi-tropical — good frost resistance.
Aspect Full sun to dappled sunlight.
Water Moderate — resents prolonged dryness.
Feeding Light dressing of blood and bone or organic plant food in spring.
Pruning May be pruned back to about two-thirds after flowering.
Propagation Cuttings usually in early summer.
Possible Problems Usually trouble free.

Backhousia citriodora

FAMILY MYRTACEAE

A small tree with bright green elliptical to lance-shaped leaves to 10 cm long. These have a delicious and refreshing lemon aroma when crushed. During summer, the tree is almost covered with white flowers clustered into rounded umbels which appear fluffy due to the numerous stamens. The stamens are backed by four petals and four calyx lobes, which remain for many months and suggest clusters of small, starry green flowers.

Origin Rainforest of Qld and NSW.
Size 8–10 m and 2–3 m wide. Slow growing, but flowering from about 2 m high.
Soil Prefers well-composted loam but adapts to most well-drained soils.
Climate All but cold, frosty districts and dry inland.
Aspect Sun to half shade. Some protection from cold winds.
Water Prefers occasional soakings during hot summers.
Feeding Organic, slow release or 'complete' fertilisers applied as flowers finish.
Pruning Not necessary, but young trees may be cut back lightly for bushier growth.
Propagation Cuttings or seed.
Possible Problems Rarely affected.

Backhousia citriodora

Astartea fascicularis

Anthocercis littorea

Baeckea brevifolia

Baeckea linifolia

Baeckea utilis

Baeckea brevifolia

FAMILY MYRTACEAE

An unusual little shrub with short bronze scale-like leaves lying along the twiggy branches which are often reddish towards tips. Attractively simple, round petalled, soft pink or white flowers appear from the leaf axils during spring and summer.

Origin NSW in moist heathland country.
Size 0.5–1 m high and to 0.5 m wide.
Soil Occurs mainly in sandstone but adapts to moist loams.
Climate Cool and temperate
Aspect Best in lightly broken sunlight with some protection from other light shrubbery.
Water Liberal to moderate — resents complete dryness.
Feeding Sparingly if at all. Compost or well-decayed animal manure is acceptable.
Pruning Prune back to about one-third after flowering.
Propagation Seed or cuttings.
Possible Problems Tea tree scale. Resents long periods of dryness.

Baeckea linifolia

Weeping Baeckea

FAMILY MYRTACEAE

A graceful small to medium shrub with deep green or sometimes bronze needle-like leaves and pendulous outer branches. Tea-tree-like white flowers to 5 mm across appear in leaf axils of outer branches, mainly throughout summer.

Origin Eastern Australian States, mainly in moist gullies along creeks, etc.
Size 2–3 m and to 2 m wide.
Soil Adapts to both sandstone or clayey soils, preferably well composted.
Climate Cooler temperate to semi-tropical.
Aspect Light shade to full sunlight.
Water Liberal — resents long periods of dryness.
Feeding Slow release or organic plant food in moderation.
Pruning Pruning after flowering results in denser growth.
Propagation Seed or cuttings.
Possible Problems Except for occasional tip-webbing caterpillar it is fairly trouble free.

Baeckea ramosissima

FAMILY MYRTACEAE

A small tiny foliaged twiggy but attractive shrub. Small tea-tree-type flowers with five rounded white or sometimes pink petals almost cover the plant in spring with scattered flowers later.

Origin All states except WA and NT.
Size Variable — more commonly about 1 m high and as wide but prostrate forms also occur.
Soil Best in well-drained gritty sandy loam. Maintain a leafy mulch around plants.

Baeckea ramosissima

Climate Cool to semi-tropical. Good frost resistance.
Aspect At least half sun.
Water Ample especially during summer, allowing surface to dry out between watering.
Pruning Prune lightly as flowering finishes.
Propagation Cuttings.
Possible Problems Check occasionally for leaf-webbing caterpillars.

Baeckea utilis

Mountain Baeckea

FAMILY MYRTACEAE

A small shrub with comparatively long rigid cane-like branches clad in short twigs carrying short heath-like foliage, and during summer, masses of single stemmed white flowers 6–7 mm across.
Origin Mainly the higher regions of NSW, Vic. and Tas.
Size 0.5–1 m high and to twice as wide.
Soil Peaty sand or light loam, preferably on the moist side.
Climate Flowers best in cool to cold areas but also worth trying in temperate areas.
Aspect Lightly broken sunlight to light shade.
Water Prefers ample water during summer — accepts fairly wet soil.
Feeding Slow release or organic plant food.
Pruning Leafy stems may be pruned back by one-third to one-half after flowering.
Propagation Seed or cuttings.
Possible Problems Usually trouble free.

Banksia

A genus embracing about 50 species exclusive to Australia except one, *Banksia dentata*, which also occurs in New Guinea. Like its close relatives, grevillea and hakea, the individual banksia flowers are in pairs and with long prominent styles, but are crowded into dense, cylindrical cone-like spikes which persist after flowering finishes to carry the woody seed capsules. The exception is *B. dryandroides*, which has globular heads of flowers. About forty of the species occur in various parts of WA, with only ten in eastern Australia, but they seem more adaptable to most parts of Australia.

Banksia coccinea

Banksia coccinea

Albany or Scarlet Banksia

FAMILY PROTEACEAE

A very showy banksia often featured in cut flower arrangements. Variable in growth, from one- to three-stemmed, 2 m high shrub to impressive many branched 4 m specimens. Leaves broad oval to oblong 6–10 cm long, lightly toothed and leathery, branches with terminal, squat cylindrical heads of scarlet flowers. These are vertically striped silver-grey while pistils of florets are still furled — appearing in spring.

Origin Southern WA particularly Albany area.
Size See general description.
Soil Well-composted slightly alkaline sandy to medium loam. Limestone chip has proved successful when added to some soils. Perfect drainage is usually specified but they are often found growing naturally in fairly moist soils.
Climate Temperate; unhappy with hot humid conditions.
Aspect Lightly broken to about two-thirds sunlight.
Water Moderate, at least until established.
Feeding Only sparingly. Slow release fertiliser may be used. Avoid high phosphorus fertiliser. Well decomposed compost or cow manure is safest.
Pruning encourage bushier growth by pruning back flower stems to leave at least four or five leaves at base.
Propagation Seed.
Possible Problems Root rot, especially in warmer moist east coast conditions.

Banksia ericifolia

Banksia ericifolia

Heath Banksia

FAMILY PROTEACEAE

A dense, dome-shaped shrub with branchlets heavily clad in dark-green, slender heath-like leaves about 2 cm long. From mid-autumn into early winter, it is handsomely decked with erect cylindrical spikes, to 30 cm long, of tightly packed coppery orange flowers. Those with unopened flowers display neat, vertical rows of hooked glossy bronze wire-like styles. An adaptable large shrub for most climates, including seaside and inland areas.

Origin East coast, mountains and western slopes of NSW, mainly in sandstone regions.
Size 2–3 m high and similar width.
Soil Indigenous to acid sandstone areas, but accepts neutral to slightly alkaline soils; also heavier loams if well-mulched with leafmould. Good drainage is essential.
Climate Cool to tropical; good frost resistance once established.
Aspect Full sun and wind exposure to dappled sunlight. Good resistance to salt winds.
Water Moderate.
Feeding Leafmould mulch is sufficient; optional light scattering of organic plant food in spring.
Pruning Not necessary, but trim in spring to reduce size.
Propagation Seed or tip cuttings.
Possible Problems Rarely affected.

'Giant Candles', with cones to 40 cm long and slightly longer foliage, regarded as a hybrid between the above and *Banksia spinulosa.*

Banksia grandis

FAMILY PROTEACEAE

Renowned for its spectacular saw-toothed foliage and large yellow cones. Varies from large shrub to medium size spreading tree. The leathery dark green leaves are cut saw-tooth fashion to the

Banksia ericifolia (typical growth)

midrib and grow to 30 cm long. The upright cylindrical yellow flowers are a similar length, appearing mainly in spring.

Origin Mainly the Perth and Darling Range area of WA.
Size 2–10 m high and 3–6 m wide.
Soil Occurs particularly in gravelly slightly limey soils. Well-drained acid sandy to light loams with limestone chip added may be successful.
Climate Temperate.
Aspect At least half sun.
Water Moderately until established; resents frequently wet soil.
Feeding As for *B. coccinea*.
Pruning Prune only lightly after flowering.
Propagation Seed.
Possible problems Root rot, especially under humid east coast conditions.

Banksia hookerana

Acorn Banksia
FAMILY PROTEACEAE

A showy banksia with slender, evenly saw-toothed leaves from 10–20 cm long forming an attractive base for the 8–9 cm long acorn-shaped flower cones. These at first are silvery velvet, then, from the base up, become overlaid with bright coppery-yellow as the styles protrude with the maturing of flowers. Flowers from winter to late spring.

Origin Coastal WA, north of Perth, mainly the sand plain for about 350 km north of Darling Range.
Size 1.5–2.5 m high and as wide.
Soil Prefers slightly limey sandy loam with good drainage. Grows well on Adelaide limestone plain. Worth trying with limestone chips added to root area before planting.
Climate Cooler temperate to semi-tropical but seems unhappy in humid summer rainfall areas.
Aspect Best in full sun with good air circulation.
Water Accustomed to fairly dry summer areas; moderate only.
Feeding Very sparingly with organic plant food, if at all. Occasional seaweed extract seems suitable.
Pruning Prune off old flower spikes.
Propagation Seed sown during autumn.
Possible Problems Root rots may be troublesome in some districts.

Banksia hookerana

Banksia grandis

Banksia grandis (typical growth)

Banksia integrifolia

Tree Honeysuckle

FAMILY PROTEACEAE

A small tree with excellent salt spray resistance. The leathery 10 cm long slender wedge-shaped to oblong leaves are dark green above, grey beneath. The 12–15 cm cone-like spikes of lime green to yellow flowers are produced almost throughout the year and much loved by Wattlebirds and other Honeyeaters. An excellent tree for front line planting to protect less resistant plants from salty winds.

Origin Vic., NSW and Queensland coastal districts.
Size To 15 m high and 4 m wide in protected areas, smaller in windswept aspects.
Soil Adapts to sandy or moderately heavy loams with reasonable drainage.
Climate Cool temperate to tropical; good frost resistance.
Aspect Fully exposed or sheltered position.
Water Moderate.
Feeding Not normally needed. Once established a light sprinkling of organic fertilisers or watering with seaweed extract may speed growth.
Pruning Not necessary. Tip prune when young if bushier growth preferred.
Propagation Seed, which in this species is shed soon after cones mature.
Possible Problems Rarely affected.

Banksia marginata

Banksia marginata

FAMILY PROTEACEAE

A variable shrub, or occasionally a rather sparse tree. The slender leathery leaves may vary from 8–12 cm long, sometimes lightly toothed but usually with edges slightly rolled under and dark green above and grey beneath. The cylindrical flower spikes, 9 cm long, appear from late winter into summer.

Origin SA, Vic., Tas., ACT and NSW, both coastal and inland.
Size Usually a low-branching shrub 2–4 m high and nearly as wide, but occasionally with sparse growth 5–7 m high.
Soil Adapts to most well-drained soils. Lighten heavy soils with leaf mulch.
Climate All but coldest.
Aspect Sun to dappled sunlight.
Water Moderate watering but fairly drought resistant once established.
Feeding As *Banksia integrifolia.*
Pruning May be lightly pruned in spring if necessary.
Propagation Seed.
Possible Problems Not normally affected.

Banksia robur

FAMILY PROTEACEAE

SEE PAGE 1

A low shrubby banksia that grows happily in swampy soils. The large stiff oblong to wedge-shaped leaves are to 30 cm long, serrated and attractive bronze-red when young. The cones are initially an unusual deep blue-green, lightening to lime then yellow as florets open.

Origin Qld and NSW mainly in fairly open swampy heathland areas.
Size To 2 m high, suckering from lignotuber and spreading 2–2.5 m.
Soil At home in swampy soils but also grows in average well drained garden soil.
Climate Cooler temperate to semi-tropical.

Banksia integrifolia

Banksia integrifolia

Banksia serrata

Banksia spinulosa

Aspect Full to lightly broken sunlight.
Water Liberal but survives dry periods.
Feeding Sparingly if at all with slow release or organic fertilisers.
Pruning If preferred stems may be cut down near or to ground level. They rejuvenate well when burnt to the ground by bush fires.
Propagation Seed.
Possible Problems Usually trouble free.

Banksia serrata

Old Man Banksia

FAMILY PROTEACEAE

A small to medium tree, often leaning and with twisted limbs. Trunk clad in corky, warted, purplish grey-brown bark. The tough leathery sawtoothed oblong leaves to 15 cm long are bright green above with lighter green beneath. The broad, slightly tapering, cylindrical flower spikes to 15 cm long are silver grey to lime green in bud, opening to buff, maturing to purplish brown and then aging to grey brown with protruding woody seed capsules that finally split like two broad lips or eyelids. It was these 'cones' that inspired May Gibbs 'bad banksia man' stories.

Origin Tas., Vic. and NSW, both coastal and mountain areas.
Size 5–10 m high, 3–5 m wide.
Soil Grows well in poor sand or sandstone country but also adapts to well-drained clayey loams if well-mulched with leafmould.
Climate Cool to semi-tropical.
Aspect Full sun to dappled sunlight.
Water Moderate, good drought resistance when established.
Feeding Resents any but light dressings of mild organic fertiliser. Leafy mulches should be sufficient.
Pruning Should not be necessary; rarely recovers from heavy lopping.
Propagation Seed.
Possible Problems Rarely affected except for inconsequential leaf dimpling psyllids or occasionally mildew.

Banksia serratifolia, similar with leaves more closely toothed. Flowers identical, but tree rarely exceeding 3 m in height.

Banksia spinulosa

Hairpin Banksia, Hill Banksia

FAMILY PROTEACEAE

Several forms, including the taller growing and broader leafed hill Banksia (*B. collina*). The form more commonly found in sandstone gullies from north of Sydney to Melbourne, has slender linear leaves 8–12 cm long, dark green above with slightly toothed margins recurving to partly cover the grey underside. Erect 10–20 cm long, coppery golden cones with metallic bronze hairpin-like styles appear during autumn and early winter.

Origin Eastern Aust.
Size 1.5–3 m and as wide; var. *collina* 4–5 m high, 2–4 m wide.
Soil Well-composted acid sandstone to medium loam with good drainage.
Climate Tropical to cooler temperate; fairly frost resistant.
Aspect Prefers broken sunlight; will grow well under tall trees.
Water Appreciates liberal amounts of water providing soil drains well but has reasonable drought resistance when established.
Feeding Resents standard 'complete' fertilisers or others high in phosphorus. Slow release fertiliser can be used sparingly to speed growth, otherwise just leafmould.
Pruning May be pruned back to about one-third after flowering providing branches are not denuded of foliage.
Propagation Seed.
Possible Problems Usually trouble free.

Barklya syringifolia

Gold Blossom Tree

FAMILY FABOIDEAE

Eventually a tree but in all but moist tropical districts remains for many years as a large dense shrub with 10 cm long glossy bright to deep green triangular to oval or heart shaped leaves resembling lilac (*Syringa*) as the species name implies but much more lustrous. The flowers are small, bright golden yellow densely packed in showy racemes to 24 cm long and occur over two to three weeks in mid-summer.

Origin Rainforest areas of Qld and north NSW.
Size Most temperate areas 3-5 m and 2 m wide, semi-tropical to 10 m or rainforest areas eventually to 20 m.
Soil Best in well-composted moist but well drained loam.
Climate All but cold frosty areas or dry inland.
Aspect Best with full sun.
Water Needs occasional soakings during hot dry periods.
Feeding Organic or slow release fertiliser after flowering.
Pruning In late winter or after flowering if needed.
Propagation Seed; air layers may also be successful.
Possible Problems Rarely affected but check occasionally for scale.

Barklya syringifolia

Bauera rubioides

Dog Rose, River Rose

FAMILY BAUERACEAE

A shrub with slender and straight branch segments clustered with whorls of stemless elliptical dark green leaflets and saucer shaped pink flowers to 2 cm across, pendant from leaf axils, most prolific during spring and early summer. A white form is sometimes available.

Bauera rubioides

Origin Eastern Australia and Kangaroo Island, frequent along the banks of coastal creeks.
Size In shaded gullies growth sparse to about 2 m and as wide. Compact to about 1 m high and wide in more open sunnier situations.
Soil At home in both light and heavy moist, but not stagnantly soggy soils, or in average garden conditions.
Climate Cool temperate to semi-tropical.
Aspect Preferably at least half sun.
Water Prefers ample water but adapts to dry periods.
Feeding Blood and bone or complete plant food in spring.
Pruning Prune back to at least two-thirds as flowers finish. Tip prune new growth.
Propagation Tip or heel cuttings.
Possible Problems Usually trouble free.

Bauera sessifolia

Bauera sessiflora

FAMILY BAUERACEAE

Unlike *B. rubioides* this species has erect growth with similar, but usually closer set, whorls of foliage. Deep rosy mauve flowers are densely clustered in leaf axils, outfacing rather than pendant, mainly during spring and summer.

Origin Victoria, Grampians area.
Size 1-1.5 m high, 0.75-1 m across.
Soil Prefers gritty loam but adapts to most well-drained soils.
Climate Cool to warmer temperate.
Aspect Good plant for light shade or full sun.
Water Responds to plenty of water with good drainage but accepts dry periods.
Feeding As *B. rubioides*.
Pruning Prune back to about two-thirds after flowering, tip prune when young, and later for more compact form.

Pruning Prune back to about two-thirds as flowers finish.
Propagation Cuttings.
Possible Problems Growth may be poor on east coast — note Soil and Water remarks. Some other varieties especially *B. purpurea*, a small shrub less than 1 m high with deep reddish purple bottle-brush-type flowers and *B. sparsa*, with larger brushes of vermilion flowers, are more adaptable.

Billardiera cymosa

Apple Berry

FAMILY PITTOSPORACEAE

A compact cushion of twining stems or, where support is available, a small climber. Young growth is silky with fine hairs, mature leaves are dull green and slenderly lance shaped. The tubular flowers flare widely at the tips to about 2 cm across, and vary from rosy mauve to light purple, lime, and mauve-blue. Flowers occur mainly from early spring to early summer and are followed by bronze-green oblong berries about 2 cm long.

Origin Vic and SA, mainly in dry mallee areas. The one illustrated is from Eyre Peninsula.
Size With support to about 2 m, otherwise 0.5–0.75m and to 1m wide.
Soil Almost any except very acid soils providing drainage is good.
Climate Temperate.
Aspect Best in open sunny situations.
Water Moderate until established, then drought resistant.
Feeding Light application of organic or slow release fertiliser during autumn or winter, providing soil is moist.
Propagation Seed or cuttings.
Possible Problems Usually trouble free.

Beaufortia squarrosa

Beaufortia squarrosa

FAMILY MYRTACEAE

A rounded but open shrub with small stiff oval and stem-clasping leaves neatly arranged in four opposite rows. Showy clusters of bright orange-red flowers with long branched styles. These appear from spring to late summer.

Origin South Western Australia.
Size Variable from about 0.6–3 m and as wide.
Soil Well drained sandy or gravelly soils but seems to resent acidity of most east coast loams.
Climate Cool temperate-to semi-tropical.
Aspect Full to half sun or lightly broken sunlight.
Water Moderate; seems to resent wet summers.
Feeding Blood and bone, or native plant food in spring.

Billardiera cymosa

Blandfordia nobilis

Billardiera scandens

Blandfordia grandiflora

Billardiera scandens

Apple Berry, Pudding Vine

FAMILY PITTOSPORACEAE

A gentle but vigorous wiry stemmed twiner with wavy margins, and linear to lance-shaped bright green leaves to 5 cm long; young shoots hairy. Greenish yellow slender bell-shaped flowers 2-2.5 cm.long are pendant from the thin stems during most of the year. These are followed by oblong edible berries.

Origin All States except WA and NT, mostly in moist, shaded gullies but also on sparsely timbered tablelands and slopes.
Size Climbs 3-5 m or in open spaces twines into a squat bush.
Soil Adapts to sand or heavy loam if well drained.
Climate Cool to semi-tropical.
Aspect Light shade to full sunlight; seems to prefer dappled sunlight.
Water Liberal to moderate; appreciates moisture but withstands dryness.
Feeding Slowly soluble, organic or sparingly applied complete fertilisers.
Pruning Not necessary. Tip pruning encourages lower branching.
Propagation Seed which is usually slow germinating.
Possible Problems Usually trouble free except perhaps for an occasional leaf-eating caterpillar.

Blandfordia grandiflora

Christmas Bell

FAMILY LILIACEAE

Small clumps of reed-like foliage to 50 cm long and erect spikes of a similar height, with a cluster of waxy red and yellow tipped bell-shaped flowers appearing mostly in December.

Origin NSW, particularly the central coast and up to the Hastings River area.
Size To about 50 cm high; base of clumps to about 12 cm.
Soil Prefer moist peaty sand or heath areas with a high watertable.
Climate Temperate to semi-tropical.
Aspect Full sun or lightly broken sunlight.
Water Ample; they need at least a moist subsoil.
Feeding Light dressing of complete plant food or fowl manure in spring and early autumn.
Pruning Remove flower heads as colour fades.
Propagation Seed; slow to show germination, flowering usually commencing the third year from sowing.
Possible Problems Usually trouble free.

Blandfordia nobilis

Christmas Bell

FAMILY LILIACEAE

Similar to *B. grandiflora* but a little smaller and tubular in shape rather than flaring at the tips into bell shape. Flowers appear during early summer.

Origin NSW coastal sandstone areas.
Soil Prefers deep sandy soil with plenty of leafmould.
Climate Temperate.
Aspect Lightly broken sunlight.
Water Moderate; more drought resistant than *B. grandiflora*.
Feeding As *B. grandiflora*.
Propagation As *B. grandiflora*.
Possible Problems As *B. grandiflora*.
Other Species *B. cunningham*, a large flowered species from Blue Mountains, NSW; *B. punicea* from Tas. with stout spikes to 1 m high, bell-shaped flowers in a larger loose spike rather than tip cluster and heavier and broader base foliage. *B. punicea* needs ample moisture and reasonable drainage.

Boronia

About 70 species, some having several different varieties. All are evergreen shrubs with opposite leaves. The latter are occasionally single but usually pinnate. Flowers have four petals, four sepals and eight stamens.
Note: *Zieria* looks similar but has four petals and only four stamens, *Eriostemon* and *Crowea* five petals and ten stamens.

Like all members of the *Rutaceae* family the leaves contain oil glands which can usually be distinguished as small translucent dots when viewed against the light. Also the flowers are hypogynous (the sepals, petals and stamens all arise from below the ovary, in this case from a small disc below the latter).

Boronia fraseri

Boronia denticulata

FAMILY RUTACEAE

A small erect, rather twiggy shrub, the narrow, light-green leaves are 2-4 cm long and lightly toothed. Starry pale-pink flowers are liberally scattered over the bush from mid-winter to late spring.

Origin Southern WA.
Size 1 m high and 50-70 cm wide.

Boronia denticula

Boronia floribunda

Soil Well-drained, preferably gritty to sandy, loam. Lighten heavier soils with rotted leafmould and a surface mulch.
Climate Cooler temperate to semi-tropical.
Aspect Accepts full sun but in warm districts prefers lightly dappled sunlight.
Water Moderate; needed most during mid- to late spring.
Feeding Mulch with leafmould and an optional light dressing of organic plant food during mid spring.
Pruning One of the few boronias that does not respond well to heavy pruning.
Propagation Seed or cuttings.
Possible Problems Rarely affected.

Boronia floribunda

FAMILY RUTACEAE

An outstanding boronia with similarly shaped but usually brighter green foliage than *B. pinnata*. It is also more floriferous, pale pink rather than rose and its growth is generally bushier. Flowers early to mid-spring.

Origin NSW, especially southern coastal and tableland areas.
Size 1–1.5 m high, 60–70 cm wide.
Soil Sandy to medium, not too dry, but well-drained loam. Heavier soils should be lightened with leafmould; maintain a surface mulch.
Climate Cool to semi-tropical.
Aspect Does well in dappled shade or morning sun.
Water Moderately, particularly during spring and summer although well established plants have drought resistance.
Feeding Light sprinkling of blood and bone or native plant manure in spring.
Pruning Prune back outer flower carrying stems by about two-thirds as flowers finish, preferably without denuding them of foliage.
Propagation Tip cuttings during late spring or scarified seed.
Possible Problems Check new tip growth for leaf-webbing caterpillars. If present squash tip and remove, which usually acts as beneficial tip pruning.

Boronia fraseri

FAMILY RUTACEAE

Spreading shrub with fairly dense slightly glossy leathery pinnate foliage, the individual leaflets comparatively broad, oval to elliptical shape. Deep pink starry flowers in spring.

Origin NSW, mostly coastal.
Size To 2 m high, 1–1.5 m wide.
Soil Prefers fairly moist but well-drained loamy soil.
Climate Temperate to cool with good frost resistance.
Aspect Prefers lightly dappled sunlight but suits all but dense shade.
Water Needs moderate watering, especially during spring and summer.
Feeding Blood and bone or native plant food in moderation during spring.
Pruning Prune back as flowers finish without denuding stems of foliage.
Propagation Scarified seed or cuttings during early summer.
Possible Problems Rarely affected but check for foliage mottling by red spider during very dry conditions.

Boronia heterophylla

Red Boronia

FAMILY RUTACEAE

Undoubtedly the showiest boronia when well grown. Growth is erect, but dense, massed with long sprays of large rosy red, fragrant, acorn- rather than star-shaped flowers from late winter to mid spring.

Origin Southern WA, usually in peaty soil continuously moistened from nearby swamps.
Size 1.5 to 1.75 m high and nearly as wide.
Soil Reasonably well-drained, but moist and well-mulched. Mix one third moistened peatmoss with quick drying sandy soils, or lighten clayey loams with compost and coarse sand.
Climate Cool to warmer temperate.
Aspect Accepts full sun but dappled sunlight is preferable during the heat of summer.
Water Water liberally. Responds well to trickle irrigation.
Feeding A handful of blood and bone or other organic plant food applied soon after flowering is beneficial.
Pruning Prune back by about one third as flowering finishes.
Propagation Seed or tip cuttings of maturing new growth.
Possible Problems Check for leaf-shoot-webbing- or leaf-rolling caterpillars, especially new growth.

Boronia ledifolia

Sydney or Winter Boronia

FAMILY RUTACEAE

A small and usually sparse shrub liberally decked with wide, open, flat, starry pink flowers during winter. The sprays are a dusky salmon-rose for several weeks after which the petals fold back to the bud formation. Leaves may be single to about 3 cm long or pinnate with 3–5 or occasionally 7 lance-shaped, to oblong-shaped, slightly rough, deep, dull-green leaflets, which always release a pleasant, strong, musky aroma when brushed or agitated.

Origin Exposed hillsides and ridges, mainly in Hawkesbury sandstone areas of coastal NSW.
Size 0.75-1 m high, 0.5–0.75 m wide.
Soil Prefers a lightly acid sandy soil with a surface mulch of leafmould. Good drainage is essential.
Climate Cooler temperate to semi-tropical.
Aspect Dappled sunlight to full sun. Ideally, light overhead foliage which gives some protection during the hottest part of summer but allows almost full sun penetration at other times.
Water Occasional soakings during dry periods. Good drought resistance once established.
Feeding Light application of blood and bone or other organic plant food during spring. Maintain at least a light mulch of leafmould.
Pruning Prune back by ⅓ to ½ immediately after flowering.
Propagation Scarified seed or tip cuttings during spring.
Pests & Diseases Not usually affected.

Boronia heterophylla

Boronia heterophylla

Boronia ledifolia

Boronia microphylla

Boronia muelleri

Boronia mollis 'Lorne Pride'

Boronia megastigma

Boronia megastigma

Brown or Scented Boronia

FAMILY RUTACEAE

The boronia best known and sought after for its sweet fragrance. A slender shrub with leaves divided at the base into small light to bright green heath-like leaflets to 2 cm long. The tips of slender outer branches are pendulous with cupped flowers varying from light tan to dark purplish brown on the backs of petals and lemon yellow inside. There is also an all yellow form usually sold as Boronia megastigma 'Lutea'. This boronia is often short lived when away from its natural environment but is worth growing even for one season's enjoyment. It flowers the second year from seed.

Origin Moist soils of southern WA.
Size Under good conditions to 1–5 m high

and 0.75 m wide but rarely more than half this size in most gardens.
Soil Well composted or peaty sandy loam that does not completely dry out but is well drained.
Climate Cool to temperate.
Aspect Morning sun or at least some protection from full sun in summer; also shelter from strong winds.
Water Liberal moisture to keep soil damp but not soggy.
Feeding Half normal strength soluble fertiliser, seaweed extract or light dressing with organic plant food in spring. Mulch to keep soil cool and moist during summer.
Pruning Pruning back to about half as flowers finish helps to prolong life of plant.
Propagation Seed or tip cuttings.
Possible problems Red spiders mites may mottle and dull foliage, especially where plants have overhead protection from weather.

Boronia mollis

FAMILY RUTACEAE

Usually a tall growing arching shrub, the pinnate foliage slightly hairy, individual leaflets elliptical about 3 cm long with pungent aroma. 'Lorne Pride' is a recently found compact form that is now available.

Origin Mainly the eastern escarpment of NSW. Plentiful in Katandra Bushland Reserve., 'Lorne Pride' is from Manning River area.
Size 2–3 m high and as wide in shaded gullies, 'Lorne Pride' to 1.5 m and as wide.
Soil Prefers well-mulched light to sandy loam, but adaptable.
Climate Semi-tropical to cool with good frost resistance.
Aspect Sun to light shade; see description above.
Water Prefers moderate moisture especially during spring to summer.
Feeding As *B. fraseri.*
Pruning As *B. fraseri.*
Propagation: Cuttings during early summer or scarified seed.
Possible Problems: As *B. floribunda.*

Boronia microphylla

FAMILY RUTACEAE

A densely compact small shrub with fine glossy pinnate foliage, the tiny leaflets wedge shaped (obovate). The numerous, small, slightly cupped flowers have comparatively broad soft pink petals, each with a deeper rose-pink centre ray and back. Flowers from early to late spring.

Origin NSW, particularly the Blue Mountains.
Size 50–90 cm high and nearly as wide but varying from prostrate to slightly taller in some locations.
Soil Sandstone soils with overlay of leaf-mould, but adapts reasonably well to all but toughest well-drained soil.
Climate Semi-tropical to cool; good frost resistance.
Aspect Lightly broken sunlight to about ⅔ sun, preferably with some protection during hottest part of day.
Water Like regular watering, especially from mid spring to mid summer.
Feeding Light dressing of blood and bone or complete plant food in spring. A leafy surface mulch is appreciated.
Pruning May be pruned back by about ⅓ after flowering.
Propagation Early to mid summer cuttings or scarified seed.
Possible Problems Rarely affected.

Boronia muelleri

Forest Boronia

FAMILY RUTACEAE

A beautiful boronia with masses of starry, pale-pink flowers during mid to late spring. The bright green fern-like pinnate leaves have a deliciously spicy fragrance.

Origin Relatively moist forest areas of Vic., especially East Gippsland
Size 1–5 m and nearly as wide but occasionally to 7 m tall in moist shaded gullies.
Soil Well composted moist, but well-drained, loam.
Climate Cool to temperate.
Aspect Prefers dappled shade and shelter from strong winds.
Water Give occasional soakings during hot dry periods.
Feeding Keep well mulched. A light sprinkling of organic plant food applied over the mulch is beneficial after flowering.
Pruning Prune back to about ⅓ as flowers finish.
Propagation Tip cuttings during early summer.
Possible Problems Occasionally leaf rolling caterpillars may web and eat out new growth tips. However this can have a beneficial tip pruning effect.

Boronia nana var. *hyssopifolia*

Boronia nana var. *hyssopifolia*

Dwarf Boronia

FAMILY RUTACEAE

A dwarf dense compact little shrub, unlike the species (*B. nana*) it has simple lance-shaped to linear rather than compound leaves. During spring it is spangled with rosy red buds which open to starry white or blush pink flowers about 1 cm across.

Origin SA, Tas., Vic. and NSW.
Size 0.25–1 m high, to 0.75 m wide.
Soil Well-composted peaty sand to medium loam, moist but with good drainage.
Climate Cool to temperate.
Aspect From light shade to full sun; some light protection is desirable in warmer areas.
Water Moderate to liberal.
Feeding Light dressing of organic or slow release fertiliser in spring as flowering diminishes.
Pruning Prune lightly after flowering.
Propagation Cuttings or scarified seed.
Possible Problems Except for possible tip-webbing caterpillar it is generally trouble free.

Boronia pinnata

Boronia pulchella

Boronia pinnata

FAMILY RUTACEAE

A lovely boronia with deep to bright glossy green pinnate foliage and slender leaflets deliciously fragrant when brushed or bruised. The clusters of partly cupped, waxy petalled flowers occur in early to mid-spring and are usually deep rose pink but there are also paler, white and multi-petalled forms.

Origin NSW mainly in the eastern areas.
Size Stems may elongate to 2 m or more in valleys but plants are more compact in open areas and with pruning. More compact forms, spreading about 1.5 m are available.

Soil Sand to medium loam providing latter is lightened by leafy mulches.
Climate Cool to semi-tropical. Fairly frost resistant.
Aspect Adaptable. Although found in moist, shaded gullies, it also occurs within range of salt spray on coast. In the latter case it is usually lower growing with darker flowers and bronze green foliage.
Water Appreciates moderately moist but well-drained soil.
Feeding As *B. microphylla*.
Pruning Prune back to about half size when flowers finish.
Propagation Sacrified seed or cuttings.
Possible Problems As *B. heterophylla*.

Boronia pulchella

FAMILY RUTACEAE

A very beautiful and showy boronia, low spreading and compact with pinnate leaves to 3 cm long made up of five to nine slender leaflets, almost entirely hidden during spring beneath a spectacular canopy of broad but wide open brilliant pink flowers about 1.5 cm across.

Origin Southern WA.
Size 0.5–1 m high, 1–1.5 m wide.
Soil Reasonably moist, well-composted light sandy to medium loam that is well drained and not too acid.
Climates Temperate; moderate frost resistance.
Aspect Usually found in lightly broken sunlight but grows well in the open rock garden at National Botanic Gardens, Canberra.
Water Moderate to liberal. Resents dryness for long periods.
Feeding Organic or slow release fertiliser in spring.
Pruning Prune back to about one-third as flowers finish.
Propagation Cuttings.
Possible Problems Usually trouble free.

Boronia serrulata

Native Rose

FAMILY RUTACEAE

A delightful and much prized boronia; the erect stems, usually in candelabra form, are clustered with serrated rhomboid shaped bright to bluish green leaves, usually all pointing upwards towards their showy crests of bright pink, cupped, fragrant flowers.

Origin NSW, mostly on slightly raised sections of fairly open sandstone gullies.
Size Usually 50–70 cm and as wide, but occasionally to 1 m.
Soil Prefers sandy loam with leafy surface mulch. Pockets of heavier loams may be lightened with coarse sand and rotted leafmould.
Climate Cool to temperate; reasonable frost resistance.
Aspect Prefers lightly broken sunlight with some protection from strong winds.
Water Responds to frequent watering during spring and summer providing soil does not remain soggy.
Feeding Blood and bone or native plant food lightly sprinkled over the leafy surface mulch in spring and watered in can be beneficial.
Pruning Prune back to at least half as flowers finish and remove the oldest twiggy branchlets entirely, providing some foliage is left on most stems.
Possible Problems As *B. floribunda*

Boronia thujona

Bronze Boronia

FAMILY RUTACEAE

A comparatively tall loose shrub with foliage resembling *B. pinnata* but with more leaflets and turning bronze in winter, lightening to a duller bronze green in spring. Also the foliage lacks the same pleasant tang, closer to a dull eucalyptus aroma. The pink flowers about 2 cm across also resemble those of *B. pinnata* but the axillary clusters are lower down the stems and occur a little later in spring.

Origin NSW, mostly in moist lightly shaded gullies.

Boronia serrulata

Size In native habitats 3–4 m and 2–3 m wide but usually only about 2 m high in cultivation.
Soil Occurs in sandstone with moist leafy mulch but adapts to reasonably well-drained lighter loams.
Climate Cooler temperate to semi-tropical. Good frost tolerance.
Aspect Best in light shade or dappled sunlight.
Water Moderate to generous. Resents complete dryness.
Feeding Organic plant food or slow release fertiliser in spring.
Pruning May be pruned back to about half after flowering.
Propagation Cuttings from hardening spring growth.
Possible Problems May suddenly die out during hot dry summers.

Boronia thujona

Bossiaea dentata

Bossiaea dentata

FAMILY FABACEAE

An erect shrub with opposite glossy green arrow — to triangle-shaped leaves about 2.5 cm long, serrated at the margins, and slender pendulous dusty salmon red pea flowers 2–3 cm long, mostly during winter.

Origin WA.
Size 1–3 m high and as wide.
Soil Moist but fairly well-drained composted sand or light loam.
Climate Temperate.
Aspect Light shade or dappled sunlight but has good wind resistance.
Water Prefers ample water but accepts periods of dryness when established.
Feeding Compost is generally sufficient but slow release fertiliser used sparingly should speed growth.
Pruning Improved by light pruning after flowering.
Propagation Seed or cuttings.
Possible Problems Usually trouble free but may suffer from red spider mites in dry atmosphere with overhead cover.

Bossiaea foliosa

Leafy Bossiaea

FAMILY FABACEAE

An attractive shrub the numerous twiggy stems radiating from the dark tube-like branches. These are edged on either side with closely set small glossy rounded and slightly convexed leaves only about 4 cm across. During late spring and summer it carries a showy mass of bright clear yellow pea flowers about 1 cm long.

Origin NSW (mainly south-eastern) and Victoria.
Size 0.5–2 m high and to similar width.
Soil Sandy soil and light to medium loam, in all cases with a leafy mulch and reasonable drainage.
Climate Cold and temperate districts. Frost resistant.
Aspect Broken sunlight, preferably with some light protection by other shrubbery also some cover for root area.
Water Moderate to liberal but once established has some drought tolerance.
Feeding Light application of slow release or native plant fertiliser is acceptable as flowering finishes.
Pruning May be lightly pruned if required.
Propagation Seed or cuttings.
Possible Problems Scale, indicated by presence of black sooty mould, especially in warmer temperate areas.

Bossiaea foliosa

Bossiaea obcordata

(syn. *B. microphylla*)

Spiny or Prickly Bossiaea

FAMILY FABACEAE

A small erect but weak stemmed shrub so called because its numerous tiny grey-green leaves are obcordate (heart shaped in reverse, with the indentation at the tip). The numerous branchlets terminate in fine spines. In spring it carries masses of small yellow pea flowers with red centres.

Origin Eastern Australia especially in coastal sandstone areas.
Size 0.5–1.5 m high and about 0.5 m wide but apparently wider and denser in some areas.
Soil At home in leafy acid sandstone but adapts to most well-drained loams.
Climate Cooler temperate to semi-tropical.
Aspect Preferably light shade or broken sunlight but accepts nearly full sun.
Water Accepts liberal water but good drought resistance when established.
Feeding Leafy mulch is sufficient; slow release fertiliser if at all.
Pruning Reduce by about one-third after flowering.
Propagation Seed or cuttings.
Possible Problems As *B. foliosa*.

Bossiaea obcordata

Bossiaea scopendria

Bossiaea scopendria

Leafless Pea Flower

FAMILY PAPILIONACEAE

An unusual plant with flattened winged stems, in most cases leafless or with tiny oval leaves towards the base. Flowers clustered on either side of short leafless stems, reddish brown in bud, opening to pea-shaped flowers to about 1 cm across, standard yellow with a reddish zone towards the base and the small keel bronze-red. These appear during early to mid spring.

Origin NSW mainly in lightly shaded coastal gullies.
Size 0.6–1.5 m high, 0.5–1 m wide.
Soil Sandstone or loam with leafy mulch and good drainage.
Climate Temperate to semi-tropical.
Aspect Lightly broken sunlight to full sun with some root cover.
Water Moderate, at least until established.
Feeding Light dressing of complete organic fertiliser in spring, or just a leafy mulch.
Pruning May be lightly pruned after flowering.
Propagation Scarified seed or cuttings.
Possible Problems Usually trouble free.

Bottlebrush

See *Callistemon*

Brachycome iberidifolia

Swan River Daisy

FAMILY ASTERACEAE

An appealing long flowering plant usually treated as an annual, with clumps of finely divided foliage below slender stemmed daisy flowers about 2 cm in diameter, in blue, mauve, purple and occasionally white, mostly with dark centres contrasting with the row of gold tipped stamens.

Origin Southern WA.
Size 25–35 cm high and nearly as wide.
Soil Any well-drained garden soil.
Climate Cooler temperate to semi-tropical.
Aspect At least half sun. Good tolerance to seaside conditions.
Water Frequently until established then moderate.
Feeding Add a little complete plant food or decomposing fowl manure to soil before sowing or planting out. Also add lime or dolomite if soil is strongly acid.
Pruning Trim off old flower heads.
Propagation Seed sown in late autumn or, in cold districts, in late winter.
Possible Problems Usually trouble free.

Brachycome multifida

FAMILY ASTERACEAE

A compact cushion of finely divided dark green foliage below a display of fine petalled mauve or lilac blue daisies each about 2 cm across during spring and summer.

Origin Eastern Australia
Size 10–12 cm high spreading 35–45 cm.
Soil Any well drained not too tightly packed garden soil.
Climate Cool to semi-tropical. Good frost resistance.
Aspect At least half sun.
Water Moderate; especially until established.
Feeding A light scattering of complete plant food or fowl manure applied in spring is beneficial.
Pruning Trim off old flower heads.
Propagation Cuttings preferably in late autumn.
Possible Problems Usually trouble free.

Brachycome multifida

Brachycome iberidifolia

Brachycome piligarensis

Brachycome piligarensis

FAMILY ASTERACEAE

A suckering plant making dense clumps of dull green lobed foliage about 2 cm wide and 6–7 cm long, displaying comparatively large, deep rose to light purple daisies 3–4 cm across with limey yellow centres; flowers mainly during spring and summer.

Origin North-western NSW.
Size Clumps to 10 cm high and 16 cm wide, spreads by suckers.
Soil Any well-drained soil.
Climate Cool to semi-tropical.
Aspect Full sun to about half shade.
Water As *B. multifida.*
Feeding As *B. multifida.*
Pruning Remove old flower heads.
Propagation Divide clumps during winter or take suckers as cuttings from late winter to early spring.
Possible Problems Usually trouble free.

Brachychiton acerifolius

Brachychiton acerifolius

Flame Tree

FAMILY STERCULIACEAE

One of Australia's most spectacular flowering trees with large glossy green broadly lobed hand-shaped leaves to 20 cm across, then, if these become deciduous in spring, the outer branches form a cascade of waxy red bells to about 1 cm in diameter on a filament of waxy red branching stems. This display is mainly from late spring to early summer. These trees are frequently used in striking contrast with Brazil's jacarandas. Flowering performance varies with season and individual trees; in most districts only trees or sections of trees that become deciduous produce flower. A hybrid between this and the kurrajong (*B. populneus*) retains foliage, and flowers more consistently but colour is usually paler. Grafted plants will be consistent with parent.

Origin Eastern coast of Australia from Cape York to southern NSW mainly in rainforest areas.
Size To 20 m high, 4–5 m in width.
Soil Most well drained loams.
Climate Tropical and temperate, not severe frost areas.

Calandrinia balonensis

Aspect Where crown can reach good sunlight.
Water Ample during early summer; a fairly dry autumn is possibly more conducive to good flowering.
Feeding Fowl manure or native plant food as flowers finish seems beneficial.
Pruning Not appropriate.
Propagation Seed.
Possible Problems 'Bag moth' or leaf-stitching caterpillars may form colonies between webbed together leaves. Cut off colonies and destroy where possible otherwise 'leave for the birds'.

Brachysema lanceolatum

Brachysema lanceolatum

Swan River Pea

FAMILY PAPILIONACEAE

A dense shrub comprised of numerous erect stems with slender leathery foliage 5–8 cm long and about 1 cm wide, dark green above, grey beneath. Long keeled deep red pea flowers along the upper half of branches occur mainly from mid-winter to mid-spring.

Origin WA.
Size 1–1.5 m high and eventually 2 m wide.
Soil Adapts to most well-drained soils.
Climate Cooler temperate to semi-tropical.
Aspect Grows in light shade under trees but flowers better in full sun.
Water Moderate, good drought resistance when established.
Feeding A little blood and bone or native plant food in spring speeds growth.
Pruning May be pruned back to below old flowers.
Propagation Scarified seed or summer cuttings.
Possible Problems Usually trouble free.

Calandrinia balonensis

Broad-leafed Parakeelya

FAMILY PORTULACACEAE

Free flowering and showy little biennial to perennial with a basal cluster of deep dull green fleshy lance-shaped leaves from 5–12 cm long, 1–1.5 cm wide at the base and sharply tapering to the tip and centre vein in a deep groove. Slender stems to 30 cm high carry several buds with one flower open. It is saucer shaped like a single portulaca, about 3 cm across, bright rosy mauve to purple with a white zone surrounding the yellow centre stamens. Each flower lasts one day. Flowers mainly in spring then sporadic throughout the year.

Origin Central Australia and spilling into the semi-arid areas of adjoining States.
Size To 12 cm high, 35–50 cm wide.
Soil Usually found in gritty red sand but adapts to well-drained and alkaline to mildly acid loam.
Climate Tropical and temperate semi-arid districts.
Aspect Full to lightly broken sunlight.
Water Establishes quickly after rain, then very drought resisting.
Feeding Complete or slow release plant food before direct sowing seed or as flowering finishes.
Pruning Not required.
Propagation Seed.
Possible Problems Seems trouble free in dry climates.

Callicoma serratifolia

Callicoma serratifolia

Creek Wattle, Black Wattle

FAMILY CUNONIACEAE

A tall shrub often forming a clump or straight black stems. Those to about 5 cm diameter are pliable and were used to weave walls of the wattle and daub dwellings of the first settlement, hence the name wattle which was also applied to acacia which has similar flowers but is unrelated. The stiff elliptical serrated leaves to 10 cm long are dark green above, greyish underneath; and clusters of cream wattle-like flowers about 1.5 cm across appear in spring.

Origin East coast of Australia, particularly along creeks or watercourses.
Size 4–5 m high and 3–4 cm wide.
Soil Fairly moist well-composted sand or loam with reasonable drainage.
Climate Cooler temperate to tropical.
Aspect Prefers shelter in dappled sunlight but also grows in full shade or sun.
Water Preference for ample water but once established has moderate drought resistance.
Feeding Leafy mulches normally sustain but slow release or organic plant foods should speed growth.
Pruning Rarely necessary but may be cut back moderately to encourage bushiness or heavily for base suckering or clumping.
Propagation Seed.
Possible Problems Borers frequently attack older trunks. Thrips may cause some foliage mottling during dry conditions.

Callistemon

A large genus of very adaptable plants, most tolerant to heavy, wet, even swampy, soil or dry sand and even to brackish soil. Wattle birds and other honeyeaters are attracted to flowers.

Some of the *Melaleuca, Calothamnus* and even unrelated banksias in some districts are categorised under the common name bottle brush. The first two mentioned also belong to the Myrtaceae family. The main distinguishing factor of *Callistemon* is that the stamens are free or separate right down to their base, whereas those of *Melaleuca* are fused into bundles; *Calothamnus*, are joined to a central rib in feather-like formations.

There are only about twenty-five definite species but there are many cultivars and an increasing list of hybrids. However, it should be noted that hybrids and cultivars can only reproduce true to their parent plant when propagated from cuttings or other vegetative means, not from seed.

Callistemon 'Captain Cook'

FAMILY MYRTACEAE

A dwarf and very free flowering bottle brush with small to medium size pendulous brushes and comparatively small closely set foliage. It has features which suggest a miniature form of *C. viminalis*. It is long flowering, mainly during spring and autumn.

Origin Probably derived from *C. Viminalis*, eastern Australia.
Size The original type is 1–1.5 m high

Callistemon 'Captain Cook'

and 75 cm wide, but many sold under this name were from seed and therefore have given variable results.
Soil Adapts to most soils but growth is more rapid in well-composted loam.
Climate Cooler temperate to semi-tropical
Aspect Full sun assures best flowering.
Water Ample, at least until established and in spring for more rapid growth.
Feeding A little complete plant food or fowl manure in spring.
Pruning Trim off old flower heads.
Propagation From cuttings only.
Possible Problems Leaf-stripping caterpillars in webbed colonies. Scale, which may be small and difficult to detect but indicated by black sooty mould on leaves and stems.

Callistemon citrinus

(syn. *C. lanceolatus*)

Crimson Bottle Brush

FAMILY MYRTACEAE

An upright comparatively stiff shrub with lance shaped leaves 6–7 cm long, and 8–10 cm long cylindrical brushes of deep crimson flowers mainly during mid-spring and occasionally later. There is also a winter-flowering form and a white form. Older stems are clustered with woody seed capsules on unpruned bushes.

Origin Eastern Australia particularly along watercourses of coastal sandstone areas.
Size 3–4 m high and to 2–2.5 m wide.
Soil Found mainly in peaty sand but adapts to clay-loams, especially when they are frequently moist.
Climate Cool to semi-tropical
Aspect Growth more compact and flowers more prolific in at least half sun.
Water At home in constantly moist soil but has good resistance to periods of drought.
Feeding Fowl manure or light dressing of complete plant feed may be given as spring flowering flush finishes.
Pruning Prune back flowering stem, leaving three or four leaves at base, as flowers begin fading otherwise new growth follows rapidly from the end of flower brush.
Possible Problems Being a stiff leafed variety is usually trouble free but see 'Captain Cook'.

Callistemon citrinus

Callistemon 'Gawler Hybrid'

(syn. *C.* Harkness)

FAMILY MYRTACEAE

A handsome bottle brush with dense compact slightly pendant growth, and bright red flower spikes to 15 cm long mainly during spring and early summer, then usually with another small flush in autumn.

Origin Hybrid possible from *C. viminalis* of NSW and Qld origin.
Size 4–5 m tall and to 3 m wide.
Soil Adapts to most soils.
Climate Temperate to tropical; moderate frost tolerance.
Aspect Needs at least half sun to flower well.
Water Usually makes more rapid growth with plenty of moisture but withstands dryness when established.
Feeding Responds to a light dressing of complete plant food or fowl manure as the first spring flowers fade.
Pruning More compact if aging flower spikes are trimmed off.
Propagation Only cutting grown plants come true to type.
Possible Problems Check for foliage stripping caterpillars in webbed colonies.

Callistemon 'Gawler Hybrid'

Callistemon 'Hannah Ray'

Callistemon 'Hannah Ray'

FAMILY MYRTACEAE

Another fine bottle brush with *C. viminalis* parentage. Similar in growth habit to *C.* 'Gawler Hybrid', flower brushes not quite as large but in branching stems and so prolific that they almost obscure foliage.

Origin Hybrid or cultivar introduced by the late Hazelwood Bros Nursery, Epping, NSW.

Size and **cultural needs** as for *C.* 'Gawler Hybrid'.

Callistemon 'King's Park Special'

FAMILY MYRTACEAE

Similar to *C.* 'Gawler Hybrid', massed with branching stems of fairly large red brushes, the plant appearing to be a little stouter.

Origin Parentage believed to be similar to the above three cultivars. Appeared in King's Park, Perth, WA.

Size and **cultural needs** as for *C.* 'Gawler Hybrid'.

Callistemon pallidus

Callistemon pallidus

Yellow Bottle Brush

FAMILY MYRTACEAE

A fairly dense shrub with slender elliptical pale to greyish green foliage to 4 cm long and profuse silky cream to yellow brushes 8–12 cm long from mid-spring into summer and often with another smaller flush in autumn.

Origin The southern half of eastern Australia.
Size 2–3 m high and nearly as wide.
Soil Adapts to most soils.
Climate Cool to semi-tropical.
Aspect Flowering is best in full sun.
Water Prefers not to dry out for long periods during hot conditions.
Feeding Light dressing of complete plant food or fowl manure in spring.
Pruning Old flower heads may be trimmed off as they finish.
Propagation Seed.
Possible Problems Usually trouble free.

Callistemon paludosus

(pink form)

FAMILY MYRTACEAE

A rather sparse shrub with narrow lance shaped to linear light green leaves to 6

Callistemon paludosus (pink form)

cm long and fluffy flower spikes 5–7 cm wide, creamy white with a soft pink flush. Flowers in summer.

Origin SA, Tas., Vic. and NSW.
Size To 3 m high, 2 m wide.
Soil Adapts to a wide range of soils including moist sand or clay-loams.
Climate Cooler temperate to semi-tropical; withstands frost.
Aspect Full to dappled sunlight.
Water Liberal but withstands dry spells.
Feeding Complete or slow release fertiliser or animal manure in autumn.
Pruning Prune well after flowering to maintain compact form.
Propagation Cuttings.
Possible Problems Generally trouble free.

Callistemon 'Reeve's Pink'

Callistemon 'Reeve's Pink'

FAMILY MYRTACEAE

A dense fairly erect shrub with lance-shaped foliage and 9–10 cm long clear pink brushes in late spring or in cool districts during early summer. 'Mauve Mist', a seedling from 'Reeve's Pink' has mauve brushes.

Origin A seedling from *C. citrinus* selected at Reeve's Nursery.
Size To about 3 m and nearly as wide.
Requirements as for *C. citrinus*.

Callistemon salignus

White Bottle Brush,
Pink Tipped Bottle Brush

FAMILY MYRTACEAE

An adaptable and quick growing small to medium tree with papery trunk and

Callistemon salignus

Callistemon subulatus

Callistemon viminalis

almost willowy outer branches with slender foliage. Its main feature is the new growth which is silky bronze-pink, often for several months. Flowers are comparatively small creamy white and occasionally pink brushes from mid-spring to early summer.

Origin Qld, NSW and SA.
Size 5-10 m high, usually 3-4 m wide, but with age occasionally to 6 m wide.
Climate Temperate to tropical but resents severe frost.
Aspect At least half sun; accepts fairly strong winds.
Water Needs at least moderately moist soils during spring and summer to grow rapidly.
Feeding Light applications of complete soluble plant food, native plant food or fowl manure in spring will speed growth.
Propagation Seed or, if to reproduce a particular flower colour, from cutting.
Possible Problems Rarely affected.

Callistemon subulatus

FAMILY MYRTACEAE

A small attractively compact and free flowering shrub with dense blue-green linear leaves to 4 cm long and during summer carries silky deep ruby red brushes of flowers 6-8 cm long.

Origin South-eastern Vic. and NSW, mainly along river banks.
Size 1.5 m high and as wide.
Soil Adapts to a wide range of moist or well-drained soils.
Climate Cooler temperate to semi-tropical; frost tolerant.
Aspect At least half sun or lightly dappled sunlight.
Water Liberal but with tolerance to dry spells.
Feeding Most complete fertilisers or animal manure with moderation.
Pruning Prune as flowers fade for more attractively compact growth.
Propagation Seed or cuttings.
Possible Problems Usually trouble free.

Callistemon viminalis

Weeping Bottle Brush

FAMILY MYRTACEAE

A graceful large shrub or small tree. There is some variation in forms but by maturity most of the species have developed a flattish spreading top with sufficient headroom to make a good decorative shade tree. Leaves are lance shaped, pale green, 6-8 cm long and, in late spring to early summer, are almost hidden behind a brilliant cascade of bright red brushes 8-9 cm long, followed by a lesser but still worthwhile repeat flowering in autumn. These taller forms are also excellent street trees and delightful alternated with white bauhinia.

Origin NSW and Qld.
Size 5-6 m high with a top spread from 4-6 m.
Soil From sandy loam to heavy clay that has been lightened by mixing in rotted leafmould or other organic matter before planting.
Climate Cooler temperate (but not severe frost) to tropical.
Aspect Full sun to occasionally broken sunlight needed for good flower display.
Water Prefer moist summer conditions but once established adapt to drought periods.
Feeding Complete, slow release plant food or fowl manure in late spring will speed growth.
Pruning Old flower heads may be trimmed off as colour begins fading if practicable although this species does not retain the tight clusters of woody seed capsules to the same extent as *C. citrinus* and others.
Propagation From seed or if character of parent is to be duplicated, from cuttings.
Possible Problems Usually trouble free, if not see *C.* 'Captain Cook'.

Calocephalus brownii

Silver Cushion Bush,
Skeleton Bush

FAMILY ASTERACEAE

A cushion of interlaced silver branches with minute stem-clasping silver leaves. Tiny insignificant flowers form during late spring or summer. Good contrast and interest plant.

Origin Coastline of WA, SA, Tas., Vic. and parts of southern NSW.
Size 20–25 cm high, 35–50 cm wide.
Soil Well-drained gritty sand preferably with a little lime or if heavier with scree of pebbles below plant.
Climate All but humid semi-tropical to tropical areas.
Aspect Needs an exposed position with good air circulation but accepts light shade. At home with salty winds.
Water Moderate, but allow surface to dry between waterings.
Feeding Light dressing of complete plant food at planting time but avoid high nitrogen fertiliser.
Pruning Prune out dead areas from centre or below plant.
Possible Problems Short lived in warmer humid districts. Spray with Bordeaux or copper spray in warmer months after removing any dead sections.

Callostemma purpureum

Calostemma luteum

Garland Lily

FAMILY AMARYLLIDACEAE

A deciduous bulbous plant that sprouts strap-like leaves with or soon after the flower buds. Showy heads of a dozen or more slender trumpet-shaped yellow flowers are carried on stems to 50 cm high, usually soon after summer rains.

Origin Along inland rivers of SA, NSW and Qld.
Size See description.
Soil Any loamy soil; prefers limey soil but is adaptable.
Climate Cooler temperate to semi-tropical.
Aspect At least half sun.
Water At least for a month or so after foliage appears.
Feeding Light scattering of fowl manure or complete plant food when growth commences.
Pruning Not applicable although removing old flower heads can strengthen bulb.
Propagation From bulb offsets or from seed.
Possible Problems Slugs and snails can be damaging.

Calostemma purpureum

Garland Lily

FAMILY AMARYLLIDACEAE

Similar to *C. luteum* except that the flowers are deep rose or purple and

Calocephalus brownii

Calostemma luteum

seem to remain a little more tubular or with petals less flared.

Origin Found in the same localities as *C. luteum* and is often found growing with it.

General requirements and cultivation as for *C. luteum*.

Calothamnus quadrifidus

One-sided Bottle Brush

FAMILY MYRTACEAE

A spreading shrub with soft appearances created by the closely set pine-like grey-green foliage 3-4 cm long veiling the erect stems, except where punctuated by bright red feathery brushes extending for 8-10 cm. These flowers continue mainly from late spring through summer into winter.

Origin Southern areas of WA.
Size 1-2 m high, to 1.5 m wide or sometimes wider with age.
Soil At home in gravelly sand but adapts to most well-drained soils.
Climate Cooler temperate to semi-tropical.
Aspect Full sun to light shade. Accepts overhead tree cover if shade is not too dense; good wind resistance if left unstaked.
Water Moderate; good drought resistance when established.
Feeding Light dressing of complete plant food or fowl manure in spring.
Pruning May be lightly pruned above the flowers in early summer, but if straggly rejuvenate by cutting back about half the branches heavily, then the remainder when new growth emerges from the first cut. Tip prune young plants to encourage branching.
Propagation Seed or cuttings.
Possible Problems Usually trouble free but leaf-webbing caterpillars can occur.

Calothamnus quadrifidus

Calothamnus villosus

Woolly Net Bush

FAMILY MYRTACEAE

Similar to *C. quadrifidus* but leaves a little shorter (2.5 cm) and a greyer green, slightly incurved and even denser. The red flowers are also similar, but they appear a little earlier.

Origin Southern WA.
Size 1-2 m high and eventually as wide.

Culture care and propagation as *C. quadrifidus*.

Calytrix exstipulata

FAMILY MYRTACEAE

A rounded bush with tiny closely set scale-like leaves that give it the appearance of cypress, until mid-winter to early spring when it is clad in a canopy of starry, rosy purple flowers about 2 cm across.

Origin Western Qld, Northern WA, and NT, particularly north from Tennant Creek to just south of Darwin.
Size 2-3 m high, to 2 m wide
Soil Found in both clayey and sandy floodplains as well as sandstone plateaus.
Climate Warmer temperate to tropical; seems to stand frost; should adapt well to warmer coast and inland.
Aspect At least half sun but stands full exposure.
Water Mainly during summer but tolerates both prolonged wet and dry conditions.
Feeding Slow release or complete organic fertiliser in spring should benefit.
Pruning Prune back lightly after flowering.
Propagation Seed or heel cuttings.
Possible Problems Seems to be trouble free although this genus sometimes succumbs to phytophthora root rot In cultivation.

Calytrix tenuifolia

FAMILY MYRTACEAE

A small usually erect shrub; stems crowded with slender, linear to cylindrical leaves, little more than 1 cm long, and spangled with comparatively broad petalled but starry flowers 1-1.3 cm across. The flowers have saucer-shaped centres which are usually yellow. The outer half is pink, lilac or light purple. There are also all-yellow forms. All flower mainly during summer.

Origin South-western WA.
Size 1-1.5 m high and as wide.
Soil Reasonably moist but well-drained, well-composted sand or gritty loam.

Calothamnus villosus

Calytrix exstipulata

Calytrix tenuifolia

Calytrix tetragona

Climate Temperate; seems better suited to WA and SA.
Aspect Prefers lightly broken sunlight or some protection from other shrubby growth without being crowded.
Water Moderate to liberal but with surface drying out between waterings.
Feeding Slow release or organic fertiliser in spring and autumn.
Pruning Prune moderately after flowering.
Propagation Cuttings or seed; the latter should result in stronger root growth.
Possible Problems Scale, and sometimes root rot in hot humid districts.

Calytrix tetragona

FAMILY MYRTACEAE

Usually a low spreading plant with dense dark green overlapping needle-like foliage, heavily spangled with small star-shaped, five petalled flowers, more commonly white but also pink. The spring flowers fall to leave a shorter reddish calyx with five short angular lobes, the tip of each extending to a long hair-like awn.

Origin All States except NT.
Size Mostly 30 cm high and about 70 cm wide but there are more upright forms to 1.5 m.
Soil Adaptable from sandstone to medium loam if well drained.
Climate Cooler temperate to tropical.
Aspect Full to dappled sunlight.
Water Moderate.
Feeding They seem to exist happily on leafmould but blood and bone in spring is acceptable.
Pruning Rarely needed but may be pruned back as calyces age.
Propagation From cuttings.
Possible Problems Usually trouble free.

Carpobrostus glaucescens

Coastal Noon Flower

FAMILY AIZOACEAE

Carpobrostus are trailing, carpeting or sometimes mounding succulents loosely known also as pig's face and found along the temperate to tropical seacoast of most countries. This species occurs mainly along Australia's east coast. The fleshy three-sided leaves are up to 10 cm long and 1 cm wide, grey-green or deep green in more sheltered fertile areas; flowers 4–6 cm across are rosy purple with fine iridescent petals followed by cylindrical edible fleshy purplish red fruits 2–3 cm long and nearly as wide.

Carpobrostus glaucescens

Origin As above.
Size 10–15 cm high and spreading 2–3 m.
Soil Well-drained sand or loam.
Climate Cooler temperate to tropical.
Aspect Suits open sunny to lightly shaded aspect. Accepts salt spray.
Water Adaptable to frequent rainfall or drought.
Feeding Low fertility plant but accepts light applications of complete plant food.
Pruning Shorten back runners if necessary.
Propagation Plant sections as runners or cuttings which strike readily, or seed.
Possible Problems In some situations older centre growth may deteriorate with fungus diseases but plants are readily renewed from outer growths.

Cassia artemisioides

Silver Cassia
FAMILY CAESALPINIACEAE

Even without its flower this compact mound of finely divided silver-grey foliage is a wonderful complement to any garden. During late winter and spring it is studded with clusters of small cupped golden yellow flowers that keep appearing for several months.

Origin Dry inland areas of all mainland States.
Size 1–1.5 m tall (higher in sheltered situations) 1–1.5 m wide.
Soil Most well-drained soils, preferably with a little lime or dolomite added if known to be acid.
Climate All but very cold areas with severe frost.
Aspect Prefers a fairly open area with good air circulation and at least half sun.
Water Moderate until established then just occasional soakings during very dry periods. Has good drought resistance.
Feeding Compost or a light scattering of blood and bone about mid-spring or autumn.
Pruning After flowering finishes, may be pruned back heavily but without denuding branches of foliage.
Propagation Scarified seed.
Possible Problems Not usually affected but plants can be short lived in frequently moist soils.

Cassia nemophila

(syn. *Cassia eremophila*).
FAMILY CAESALPINIACEAE

A shrub similar in flowers and foliage to *C. artemisiodes* but foliage is greener, more slender and growth usually taller. However there is variation in exact detailing of the actual species because so many natural hybrids exist.

Origin As *C. artemisiodes*.
Size 2–2.5 m high, 1.5–2 m high.

Culture, care and propagation as for *C. artemisiodes*.

Cassia x *Sturtii* var. *desolata*

Cassia artemisioides

Cassia nemophila

Castanospermum australe

Celmisia asteliifolia

Cassia x *Sturtii* var. *desolata*

(syn. *Cassia desolata*)
FAMILY CAESALPINIACEAE

An appealing little cassia generally listed as *C. desolata* but considered one of several *C. sturtii* natural hybrids. Stems are light brown with blue-grey oval to obovate phyllodes 3–6 cm long and about 1 cm wide singly or sometimes in pairs liberally adorned with short racemes or clusters of five to ten buttercup-like flowers 2–3 cm across. These come mainly in spring but in semi-arid areas in spasmodic bursts after rain.

Origin NT and spilling into neighbouring States.
Size 1–1.5 m high, 1 m wide.
Soil Usually found in gravelly soil but should adapt to most well-drained alkaline to slightly acid loams.
Climate Best suited to arid or semi-arid temperate to tropical areas but appears to have reasonable adaptability and stands at least moderate frost.
Aspect Open sunny position.
Water Moderately until established; good drought resistance.
Feeding Low fertility shrub but light sprinkling with complete plant food when soil is moist in spring should promote growth.
Pruning May be pruned back to at least half after flowering.
Propagation Seed.
Possible Problems Appears fairly trouble free.

Castanospermum australe

Black Bean,
Moreton Bay Chestnut
FAMILY PAPILIONACEAE

An erect tree with lush glossy green pinnate foliage divided into elliptic leaflets 8–12 cm long, almost black trunk and branches, the latter clustered with rather open, long stamened, red and yellow pea flowers during late spring or early summer. These give way to thick woody pods 17–20 cm long that eventually open, looking like two miniature rowing boats with large chestnut-like but non-edible seeds packed in fibre resembling foam polystyrene. Now also popular as an indoor foliage plant.

Origin Qld and northern NSW rainforest areas.
Size In cultivation 8–10 m high and 4–5 m wide, but to 30 m high in rainforests.
Soil Adapts to most well-drained soils.
Climate Fairly frost-free temperate to tropical; moderate to bright light as a house plant.
Aspect Like most rainforest trees it starts its life in shade or dappled sunlight but needs at least half sun to flower. Some shelter from cold winds desirable.
Water Appreciates plenty of water during summer but once established is adaptable.
Feeding Slow release, blood and bone or complete plant food in spring.
Pruning Should not be necessary.
Propagation Seed.
Possible Problems Rarely affected.

Cat's Paw

See *Anigozanthus*

Celmisia asteliifolia

Snow Daisy
FAMILY ASTERACEAE

A small clumpy perennial with slender tapering foliage 12–25 cm long initially clad in silver-grey hairs, upper surface smooth deep green with maturity, during summer displaying white or occasionally blush pink daisies about 5 cm across on slender silver-grey stems.

Origin Alpine regions of Tas, Vic, and southern NSW.
Size 5–12 cm high and spreading to 50 cm.
Soil Very well-drained gritty sandy loam with up to one-fifth peat moss added and if very acid a sprinkling of dolomite. A surface scree is also beneficial.
Climate Cold to cooler temperate.
Aspect At least half sun and good air circulation.
Water Moderate to ample but allow surface to dry out between waterings.
Feeding Slow release or light application of complete fertiliser in spring.
Pruning Remove old flower heads.

Propagation Seed or division.
Possible Problems Loss due to dampening off fungi, especially in humid areas; surface scree may aid prevention.

Ceratopetalum gummiferum

NSW Christmas Bush
FAMILY CUNONIACEAE

A dense foliaged conical large shrub or small tree with deep green trifoliate leaves divided into three lightly toothed lance-shaped leaflets to 5 cm long. During spring the bush is cream with terminal sprays of small flowers. The fleshy calyx lobes of these gradually expand like petals to form masses of pinkish then, hopefully by Christmas, red 'flowers' about 1.5 cm across.

Origin Mostly the coastal and lower mountain sandstone regions of NSW.
Size 3–6 m high. 2–3 m wide; much taller (but leggier) in timbered gullies.
Soil Sandy or fairly deep medium loam with good drainage.
Climate Temperate to semi-tropical.
Aspect Needs at least two-thirds sunlight to flower and colour well and some shelter from very strong winds.
Water Avoid long period of dryness from late winter to late summer.
Feeding Fowl manure or complete plant food and good surface mulching during mid-summer. Do not fertilise until well established.
Pruning Prune by Christmas or just after; prune as if you were cutting good long sprays of 'bush', leaving about finger length stubs of the younger wood. Very old unattractive trees may be rejuvenated by cutting to within about 50 cm of ground during early summer.
Propagation Seed taken from parent plants that colour well.
Possible Problems Leaf rolling of new growth caused by minute psyllids but this does not seem to cause serious damage. Apart from shortage of sunlight, colour, or lack of it, is usually an inherent character, but severe dryness or thrip attack in spring can cause flowers to fall without maturing. Frequent overhead watering may then be preferable to spraying insecticides as the latter can deter flies and bees from pollinating flowers.

Chamelaucium uncinatum

Geraldton Wax
FAMILY MYRTACEAE

A spreading shrub attractive in the garden and valued for its long lasting daintiness as a cut flower. Waxy needle-like, but non-prickly, foliage. The long stemmed trusses of waxy, fine petalled flowers come mainly from late winter into mid-spring; more commonly soft pink, but cultivars range from white to rosy purple.

Origin Temperate coast of WA.
Size 2–3 m high and at least as wide.

Ceratopetalum gummiferum

Chamelaucium uncinatum

Chamelaucium uncinatum (white form)

Chamelaucium uncinatum 'Purple Pride'

Soil At home in fairly well-mulched limey sand to light loam but adapts to acid sandstone or other well-drained light loams.
Climate Cooler temperate to semi-tropical, reasonable frost resistance.
Aspect Needs at least half sun to flower well, and protection from severe winds.
Water Moderate, but accepts summer dryness.
Feeding Blood and bone, compost and complete native plant food in spring.
Pruning As you would for cut flowers but leave about 5 cm of the newer stems to carry the flower trusses; prune before flowering finishes.
Propagation From heel or tip cuttings taken in late spring or summer.
Possible Problems Reputedly short lived, but this is usually due to bad drainage or root disturbance. Start with a small plant, unstaked, and allow it to sprawl or, alternatively tip prune to encourage low branching, self bracing form.

Chorizema cordatum

WA Flame Pea

FAMILY PAPILIONACEAE

A showy erect soft wooded shrub with fairly stiff mid-green elliptical to heart-shaped leaves to 4 cm and long, mostly terminal, sprays of bright orange-red pea flowers 1–1.5 cm across with bright yellow centres and a purplish keel; flowers mainly from early to late spring.

Origin Coastal southern WA and Darling Range.
Size To 1 m high and 60 cm wide, although some forms splay to 2 m if unpruned.
Soil Preference for sandy soils but adapts to well-drained loam.
Climate All but very frosty temperate to semi-tropical areas.
Aspect Dappled shade to full sun.
Water Moderately; good drought resistance when established.
Feeding Light sprinkling of blood and bone or fowl manure in spring.
Pruning When flowers finish prune back to about half providing this leaves some foliage on most stems.
Propagation Sow scarified seed.
Possible Problems Rarely affected.

Chorizema cordatum

Clematis aristata

Clematis aristata

Traveller's Joy

FAMILY RANUNCULACEAE

A vigorous creeper with leaves divided into three broad elliptical toothed leaflets and carrying a canopy of four-petalled, creamy white, starry flowers 3–5 cm across, mainly during mid to late spring. These are followed by a fluffy mass of silver haired seed awns.

Origin Eastern Australia, in both rainforest and open country.
Size Vine may climb 10 m or more and eventually as wide.
Soil Prefers moist cool soil but reasonably well drained.
Climate Cool to semi-tropical.
Aspect Need to reach at least half sun or lightly broken sunlight for good flower display.
Water Prefers moist conditions during summer but develops drought tolerance.
Feeding Compost in spring plus a little fowl manure.
Pruning Only if necessary to contain size.
Propagation Seed or layers.
Possible Problems Usually trouble free.

Clianthus formosus

Clianthus formosus

(syn. *Clianthus dampiera*)

Sturt's Desert Pea

FAMILY PAPILIONACEAE

SA's floral emblem, and one of the most spectacular wildflowers, with prostrate stems carrying soft downy grey-green pinnate foliage divided into nine to fifteen elliptical leaflets. Erect stems 10–15 cm high carry a cluster of blood red pea flowers 6–8 cm long, more frequently with a prominent black boss in the centre. In their native desert conditions plants appear after good rains and flower within about three months, any time from late winter to autumn.

Origin Dry inland areas of all mainland States and the Fortescue district of WA.
Size Prostrate to semi-prostrate spreading 1–3 m.
Soil Accustomed to stony or sandy areas with a limestone base. In wetter areas build up beds or large containers 30 cm or more deep with gritty sand, some well-rotted compost and lime or limestone chip.
Climate Temperate to semi-tropical.
Aspect Open position with full sun.
Water Good soaking after sowing, then water when soil is dry for 3–4 cm down.
Feeding Mix in blood and bone or bone dust prior to sowing then sprinkle lightly as a side dressing when flower buds appear.
Pruning Remove old flower heads but usually treated as annuals.
Propagation Scarified seed sown direct into permanent position; germination takes 10–12 days.
Possible Problems Overwet conditions or nematodes (eelworms). Best treated as annuals.

Cochlospermum fraseri

Commersonia fraserii

Comesperma ericinum

Cochlospermum fraseri Sub. sp. *heteroneurum*

Kapok Bush

FAMILY COCHLOSPERMACEAE

An open and usually spindly small tree, leaves dark green lobed 10–15 cm across and 10–12 cm deep, deciduous during the winter/spring flowering period when spectacular heads of 5–7 cm across bright yellow flowers are carried at the ends of yellowish brown branches. These are followed by smooth green then grey-brown capsules 10–12 cm long with seeds packed amongst the downy kapok.

Origin NT and WA, towards edge of floodplains or sandstone outcrops.
Size 3–5 cm high, to 3 m wide.
Soil Well-drained sandy to medium loam.
Climate Semi-tropical to tropical.
Aspect Full sun to light shade.
Water Ample water during summer growing period and then keep on dryish side during late autumn–winter.
Feeding Accepts light dressing of complete or organic fertiliser during summer growing period.
Pruning After flowering should encourage denser growth.
Propagation Seed.
Possible Problems Rarely affected in natural habitats.

Comesperma ericinum

FAMILY POLYGALACEAE

An erect spindly shrub, sometimes only single stemmed, closely set with stiff slender leaves to 1 cm long and tapering cone-shaped heads of rosy purple buds resembling heads of wax matches, the lower ones opening to tiny bow-shaped flowers usually less than 1 cm across, the minute petals forming a tight cowl over the centre and on either side a comparatively long cupped sepal, mainly during spring and summer.

Origin Eastern States.
Size 1 m or with taller growth may straggle to 2–3 m.
Soil Mainly occurs in sandstone areas but fairly adaptable.
Climate Cooler temperate to semi-tropical.
Aspect Half sun or light shade with some wind protection.
Water Moderate.
Feeding Blood and bone in spring optional; leaf mulch should be sufficient.
Pruning Tip pruning when young would probably encourage more side branching. Prune stems to about half as flowers finish.
Propagation From side cuttings as they mature after pruning.
Possible Problems Usually trouble free.

Commersonia fraserii

FAMILY STECULIACEAE

A densely branched shrub with slender elliptical leaves 7–9 cm long, toothed margins, deep glossy green above, grey-green below and axillary clusters of flattish cream flowers about 2 cm across, resembling the classical frost crystal shape, during spring and summer. Good background screen.

Origin Eastern Australia.
Size 2–6 m high, to about 4 m wide.
Soil Most well-drained soils.
Climate Cool temperate to tropical; reasonable frost tolerance.
Aspect Accepts full shade or sun.
Water Ample to moderate until established then moderate drought resistance.
Feeding Slow release or organic food will speed growth.
Pruning For more compact growth, branches may be shortened to about one-third after flowering.
Propagation Cuttings.
Possible Problems Seems fairly trouble free.

Conospermum stoechadis

Smoke Bush, Smoke Grass

FAMILY PROTEACEAE

Shrub with numerous erect wiry stems, rigid grass-like leaves 8–16 cm long, beneath a silver-grey woolly cloud of downy little cone shaped flowers and branching stems. Unlike most Proteaceae, the flower does not display a prominent pistil.

Origin Mainly the sandplains of southern WA.
Size 1–1.3 m high and a similar width.
Soil Well-composted sand.
Climate Temperate.
Aspect Open but with some light wind protection from neighbouring, but not overshadowing, shrubbery.
Water Moderate; surface should dry between waterings.
Feeding Light composting. It seems to flourish mainly after a bushfire.
Pruning Remove old flower heads.
Propagation Seed.
Possible Problems Does not seem adaptable to average garden soils and conditions.

Conospermum stoechadis

Conostylis aculeata

Cone flower

FAMILY HAEMODORACEAE

A small clumpy plant delightful for rock gardens. Clumps of strap-like leaves to 30 cm and stems of globular clusters of deep yellow woolly buds gradually opening from the base into starry, creamy yellow flowers, from early spring into summer.

Origin Southern WA.
Size 30–35 cm high, 20–30 cm wide.
Soil Well-composted sand or well-drained light loam.
Climate Cool to semi-tropical.
Aspect At least half sun.
Water Moderate; better performance if soil is not dry for long periods.
Feeding Blood and bone, fowl manure or complete plant food during spring.
Pruning Remove old flower heads.
Propagation Seed or division in late autumn to early winter.
Possible Problems Usually trouble free.

Cordyline stricta

(syn. *Dracaena stricta*)

Palm Lily

FAMILY AGAVACEAE

A plant with dramatic form for shaded parts of the garden, forming a clump of tall canes, the upper 50–70 cm carrying rosettes of strap-like dark green leaves 30–60 cm long and to 2.5 cm wide. A

Conospermum stoechadis (natural habitat)

multitude of tiny, pale violet flowers, no more than 1 cm across are closely set along a semi-pendulous branching pannicle 30–70 cm long during mid to late summer. Flowers may be followed by purple to black berries. There is also a form with leaves widening to about 4 cm towards the centre.

Origin Qld and NSW mainly in rainforests or deep sandstone gullies.
Size 2–4 m high to 2 m wide, but often suckers more widely.
Soil Any moderately drained soil (suckers more widely in sand).
Climate Cooler temperate to semi-tropical or indoors.
Aspect Semi to full shade, preferably with frost protection or indoors.
Water Appreciates ample water but accepts long periods of dryness. Allow to dry out between waterings indoors, especially during winter.
Feeding Responds to occasional soluble plant food applications or slow release fertiliser during spring and summer.
Pruning Overleggy stems may be pruned back in spring to encourage branching from just below cut, spade off unwanted suckers.
Propagation Chiefly from finger-length sections of stem, also from seed or division of clumps.
Possible Problems Brown blotched margins may occur in container-grown plants from perlite in potting mix or fluoride in water. A tablespoon of lime water every three months should counter the latter.

Conostylis aculeata

Cordyline stricta

Correa

The flowers of *Correa*, like *Boronia*, have eight stamens and four petals. However, unlike *Boronia*, the petals of *Correa* flowers are united in a tube. This gives quite a different appearance.

Correa reflexa

Correa alba

FAMILY RUTACEAE

A dense and compact low branching shrub popular for groundcover. Stems are covered with buff-grey down, the opposite, rounded leaves to 4 cm long are dull green above, downy beneath. Flowers erect or outfacing, a short bell shape which flares to four pointed, starry petals; flowers spasmodically throughout the year.

Origin Tas., SA, Vic. and NSW mainly along seacoast.

Correa alba

Size 1.5 high, to 1 m wide; slightly taller in light shade.
Soil Well-drained preferably sandy or gritty loam.
Climate Cool to semi-tropical.
Aspect More compact and flowers better in open sunny situations but adapts to light shade. Good resistance to salty winds.
Water Moderate.
Feeding Compost, slow release fertiliser or fowl manure used sparingly to speed growth.
Pruning Rarely needed but accepts occasional trimming.
Propagation From cuttings.
Possible Problems Usually trouble free.

Correa reflexa

Common Correa

FAMILY RUTACEAE

An interesting shrub with many forms: upright or low spreading and colours ranging from green to red or red with yellowish tips. Stems are hairy and so are the leaves in most forms. Leaves are always opposite, dark green above, paler beneath but may be elliptical to heart shaped or slender oblong to 5 cm long. The gracefully pendant tubular to bell-shaped flowers are 4 cm long, partly hooded by leaf bracts and appear for a long period from late autumn to mid-spring.

Origin From Qld down through NSW, Vic. SA and southern WA, mainly from the ranges to the coast.
Size From upright 2 m shrub to prostrate about 50 cm and spreading 2–3 m; ideal for rock gardens or underplanting.
Soil Grows mainly in acid sandstone soils but adapts to well drained medium or compost lightened heavier loam.
Climate Cool to semi-tropical.

Aspect Light shade or dappled sunlight to almost full sun.
Water Moderate to liberal, especially in late spring early summer but reasonable resistance to dry periods when established.
Feeding Mulch; light scattering of blood and bone in spring.
Pruning Optional, preferably early spring if at all.
Propagation Cuttings.
Possible Problems Apart from an occasional leaf-rolling caterpillar, it is usually trouble free.

Crinum pedunculatum

Beach or Swamp Lily
FAMILY AMARYLLIDACEAE

An imposing evergreen plant with large fleshy mid-green leaves, in moist situations to over 1 m long and 15 cm wide. Leaves are several layers deep, radiating from a trunk-like bulbous penduncle to 70 cm high and about 20 cm thick. Wide fleshy stems carry a showy umbel of long, slender, funnelled, white flowers with spidery drooping petals to 10 cm long and long erect stamens.

Origin Coastal Qld and NSW around coastal swamps, behind beaches; also found on Pacific Islands as the large corky seeds will float very long distances.
Size 1.5–2 m high, to 2.5 m wide. Largest in moist situations.
Soil Prefers moist composted sand but adaptable.
Climate Temperate and tropical.
Aspect Full to half sun or light shade.
Water Takes all it can get but adapts to dry conditions when established.
Feeding Compost, manure or complete fertiliser in summer.
Pruning Remove withering under layer of foliage and old flower heads.
Propagation From the large tuber like seeds which are 5–8 cm across.
Possible Problems Slug and snail damage.

Crotalaria laburnifolia

Bird Flower
FAMILY PAPILIONACEAE

A soft wooded, quick growing shrub with soft green oval to elliptical leaflets to 7 cm long and 2.5 cm wide. The erect and tapering terminal flower racemes are to 35 cm long, with a succession of buds opening from the base into large, yellowish, green pea flowers to 6 cm long, resembling a bird in flight attached to the stem by its beak; flowers during summer and late autumn.

Origin Qld.
Size 3–4 m and 2–3 m wide.
Soil Adapts to most well-drained loam.
Climate All but very frosty areas.
Aspect At least half sun.
Water Moderate or more.
Feeding Rarely needed but a little complete plant food during spring will speed growth.
Pruning Cut back to at least half during winter or in very frosty areas in early spring.
Propagation Scarified seed.
Possible Problems Rarely affected but bushes last only about 4 or 5 years; even less in wet, cold soils.

Crinum penduculatum

Crotalaria laburnifolia

Crowea

Very similar to *Eriostemon* with flowers having five petals and ten stamens. Stamens have a minutely bearded appendage on the anthers; flowers are solitary in the leaf axils usually during summer and autumn.

Crowea exalata

FAMILY RUTACEAE

Small rounded evergreen shrub with slender club-shaped foliage to 5 cm long and scattered with small bright pink flowers 1.5–2 cm across during summer and early autumn.

Origin Eastern Vic. and Grampians; coastal belt and north-western slopes of NSW and ACT area.
Size To 80 cm high, 50–60 cm wide.
Soil Most well-composted and well-drained sands or loams.
Climate Cool to semi-tropical.
Aspect At least half sun or lightly broken sunlight.
Water Moderate.
Feeding Moderate compost or light dressings of blood and bone.
Pruning May be pruned lightly when flowers finish.
Propagation From cuttings or scarified seed.
Possible problems Rarely affected.

Crowea exalata

Crowea saligna

(syn. Eriostemon crowei)
FAMILY RUTACEAE

A beautiful little shrub with bright green slender elliptical to lance shaped leaves, 4–6 cm long; flowers are delightful, upfacing, soft to bright pink, broad petalled, to 3.5 cm across; flowers mainly from mid-summer into late autumn.

Origin NSW mainly in coastal sandstone frequently, under *Angophora costata.*
Size 0.60–1 m high and similar width.
Soil Prefers sand with leafy mulch but accepts well-drained heavier loams if lightened with compost and mulched.
Climate Cool to semi-tropical with some protection from heavy frost.
Aspect Happiest in dappled shade.
Water Moderate.
Feeding As *C. exalata.* Resents heavy feeding.
Pruning Light pruning as flowers finish optional.
Propagation Cuttings or scarified seed.
Possible Problems Usually trouble free if not overfed or excessively wet for long periods.

Cryptostylus

SEE PAGE 229

Cymbidium

SEE PAGE 222

Crowea saligna

Dampiera adepressa

Purple Beauty Bush
FAMILY GOODENIACEAE

A showy dampiera producing a dense clump of slender stems, at first erect and hairy then becoming smooth and arching with the weight of both axillary and terminal heads of flowers to 2.5 cm wide, usually pale violet to royal blue, sometimes lavender blue to purple and occurring in spring. Leaves are leathery grey-green, lance shaped to oval, 2.5–3 cm long.

Origin Qld and NSW.
Size 0.5 m high spreading to 1.5 m.
Soil Light peaty to medium loam that does not dry too rapidly but reasonably well drained.
Climate Temperate and semi-tropical.
Aspect Dappled shade to full sunlight.
Water Ample to moderate; stands short periods of dryness.
Feeding Slow release or organic fertiliser as new growth commences.
Pruning Prune back to about half after flowers finish.
Propagation Cuttings.
Possible Problems Seems to be fairly free from problems.

Dampiera diversifolia

FAMILY GOODENIACEAE

A showy, low matting or trailing plant with bright green stems and foliage, the latter mostly slender, occasionally toothed but varying in shape, and almost covered with rich royal blue flowers about 1 cm across, mainly during spring with a more scattered display into summer. A beautiful rock garden plant.

Origin Southern WA.
Size Only 3-5 cm high with a spread of about 60 cm; under good conditions spreads to over 1 m.
Soil Fairly moist but well-drained sandy or medium loam.
Climate Cooler temperate and temperate.
Aspect Lightly broken sunlight or morning sun.
Water Should not be allowed to dry out for long periods.
Feeding Sparingly; small amount of slow release fertiliser in spring.
Pruning Not applicable.
Propagation Cuttings or layers.
Possible Problems Trouble free except for slugs or snails eating flowers.

Dampiera glaberescens

FAMILY GOODENIACEAE

Appealing little clump of densely packed stems, usually the centre erect, outer sprawling, slender grey-green foliage and generously decked with about 1 cm across deep rich violet-blue flowers mainly in spring.

Origin WA.
Size 25-30 cm high and similar width.
Soil Well-drained preferably composted sand or light loam, if possible with a pebbly scree.
Climate Cooler temperate to semi-tropical.
Aspect At least half sun.
Water Moderate.
Feeding Light sprinkling of blood and bone or slow release fertiliser in spring.
Pruning May be pruned back lightly after flowering.
Propagation Cuttings.
Possible Problems Rarely affected.

Dampiera linearis

FAMILY GOODENIACEAE

A low spreading suckering plant with variable foliage, sometimes linear as the species name implies. However, leaves are frequently wedge shaped, 2-3 cm long and slightly toothed as the form illustrated and this has caused it to be sold as *D. hederafolia*. Spring and summer flowers are 1-to 2 cm across, light violet to purple, usually with a white or creamy yellow throat and with hairs on the underside. There is also a pale blue form.

Origin Southern half of WA, mainly coastal.
Size 20-60 cm high 0.5-2 m wide.
Soil Gravelly sand to crumbly well-drained loam.
Climate Cooler temperate to semi-tropical.
Aspect Half or broken to full sunlight.
Water Moderate; resents overwet conditions during summer; good drought resistance.
Pruning Trim lightly after flowering.
Propagation Cuttings or division of suckers; cut back to about one-third before replanting.
Possible Problems Usually trouble free although fungus or nematodes may shrivel foliage in moist areas.

Dampiera luteiflora

Yellow Dampiera

FAMILY GOODENIACEAE

The only yellow-flowered dampiera and quite a striking plant, closely studded with bright yellow buttercup-like flowers in small terminal pannicles during spring

Dampiera diversifolia

Dampiera linearis

Dampiera adepressa

Dampiera glaberescens

to mid-summer. Growth is light but denser with a mass of fine branches and elliptical leaves to 2.5 cm long and about 1 cm wide, carrying fine hairs on both sides.

Origin Coastal half of WA and as far east as Coolgardie.
Size 35–75 cm high and similar width.
Soil Needs well-drained light crumbly loam or gravelly sand.
Climate Cooler temperate to semi-tropical.
Aspect Dappled to full sunlight.
Water From dry summer area but is doing well in Melbourne Botanic Gardens.
Feeding As *D. linearis.*
Pruning Prune back to about half as flowers finish.
Propagation As *D. linearis.*
Possible Problems Not widely cultivated but seems trouble free.

Dampiera luteiflora

Dampiera purpurea

Dampiera purpurea

FAMILY GOODENIACEAE

A small suckering shrub resembling the broader foliaged forms of *D. linearis* but with growth more erect and open, leaves and flowers a little larger, the latter appearing more often in short racemes rather than leaf junctions.

Origin Eastern Australia.
Size 1–1.5 m high with similar width.
Climate Cool to semi-tropical.
Aspect At least half sun.
Water Responds to ample water during spring and summer but once established has reasonable drought resistance.
Feeding As *D. linearis.*
Pruning Prune back lightly after flowering.
Propagation As *D. linearis.*
Possible Problems Usually trouble free.

Dampiera stricta

FAMILY GOODENIACEAE

A sparse but worthy little plant with suckering semi-erect branching dark green stems carrying occasional leaves, mostly spatulate but varying in shape, and a scattering of bright flowers to about 15 mm across. Flowers are commonly a deep sky blue but different forms vary from white through pale to dark blue; flowers mainly from late winter into early summer. Delightful companion for *Tetratheca*, smaller boronias, etc.

Origin Widely scattered through eastern States.
Size 25–50 cm, sometimes taller with support from other plants.
Soil Adapts to most fairly well-drained soil if heavier clay-loams are loosened by leafmould addition.
Climate Cool to semi-tropical.
Aspect Occurs mostly in lightly broken sunlight but also accepts full sun.
Water Moderate
Feeding Responds to slow release or organic fertiliser and fowl manure but also accepts low fertility soils.
Pruning Cut back as flowers finish.
Propagation Cuttings.
Possible Problems Usually trouble free.

Dampiera rosmarinifolia

Wild Rosemary

FAMILY GOODENIACEAE

A dwarf suckering shrub resembling rosemary and like the latter may have erect or arching grey stems. Also the slender foliage is smooth dark green above and downy grey beneath. Liberally scattered with flowers 1.5–2 cm across, mainly during spring and early summer. Colour is usually deep blue but can be mauve-blue, pinkish mauve, purple or white.

Origin Vic. and SA.
Size 30–75 cm high, 0.5–1.5 m wide.
Soil Any reasonably well-drained. Loam or sand (not too acid).
Climate Cooler temperate to semi-tropical.
Aspect At least half sun.
Water Moderate; good drought tolerance.
Feeding Slow release or organic plant food in spring.
Pruning May be lightly trimmed after flowering.
Propagation Cuttings, preferably from suckering growth.
Possible Problems Rarely affected.

Dampiera stricta

Darwinia leptantha

Dampiera rosmarinifolia

Darwinia lejostyla

Darwinia lejostyla

FAMILY MYRTACEAE

A decorative dwarf shrub with slender, stiff glossy dark green leaves, only about 1 cm long, radiating from the numerous branches; pendulous at the tips, each branch supports a large glossy rosy purple bell 3–4 cm long and 2–2.5 cm wide. These graceful bells appear from spring to early summer. An attractive rock garden or container plant.

Origin Southern WA, especially in the Stirling Ranges.
Size 30–80 cm high, 60–80 cm wide.
Soil Any well-drained soil if not excessively compacted.
Climate Cooler temperate to temperate.
Aspect At least half sun with good air circulation.
Water Moderate.
Feeding Slow release or organic plant food in spring and again in autumn.
Pruning Prune back lightly after flowering and also when young to encourage branching and therefore more flowers.
Propagation Cuttings, as young growth is losing sappiness.
Possible Problems May be prone to fungus attack in moist sheltered situations.

Darwinia leptantha

FAMILY MYRTACEAE

A small, usually sparse shrub with grey-green heath-like opposite leaves, 1–1.5 cm long, in four rows along the stem or on older growth often confined to tufts surrounding the flowers. The latter are in tight clusters, initially white, flushed red as they age and each with a long, erect red style; flowers mostly throughout spring.

Origin NSW along temperate coast, mostly in moist heathland areas.
Size 0.5–1 m high, 30–60 cm wide.
Soil Moist peaty sand to light loam.

Climate Temperate.
Aspect Prefers lightly broken to about half sun.
Water Ample.
Feeding Compost or organic plant food in spring.
Pruning May be pruned back lightly after flowering.
Propagation Cuttings.
Possible Problems Seems trouble free.

Darwinia meeboldii

Cranbrook Bell

A delightful little shrub with stiff, dark green, 1 cm long leaves densely cladding the slender branches. Branches are initially almost erect then arch under the weight of 3 cm long, pendulous flowers like white porcelain bells with ornately pointed petals tipped rosy red. These appear in spring.

Origin Stirling Ranges of WA, in fairly moist situations.
Size 1–2 m high and to 1.5 m wide.
Soil Needs well-drained, but moist, slightly acid, light loam or peaty sand, well-mulched with compost.
Climate Temperate.
Aspect Best in dappled sunlight or morning sun.
Water Ample to moderate. Resents long periods of dryness.
Feeding Slow release fertiliser or blood and bone.
Pruning Prune lightly after flowering, reducing outer branches to as much as one-third without denuding plant of foliage.
Propagation Cuttings or layering.
Possible Problems Root damage by heavy fertilising or by cultivation disturbance; also by hot soil exposed to full sunlight.

Darwinia meeboldii

Daviesia latifolia

Daviesia latifolia

Bitter or Hop Pea
FAMILY PAPILIONACEAE

Medium shrub usually with slender arching branches and broad bluish green leathery oval to elliptical leaves to 10 cm long; outer stems carry numerous racemes, 5–7 cm long, packed with small yellow and brown, slightly fragrant pea flowers, mainly during early to late spring. The common name refers to its reported use as a hops substitute.

Origin East coast of Australia, especially in highland areas.
Size 1.5–3 m high, 1–2 m wide.
Soil Sandstone to moderately heavy well-drained loam.
Climate Cool to semi-tropical.
Aspect Generally found in light to dappled shade but also accepts full sun, especially in cool to cooler temperate areas.
Water At least moderate during spring to early summer but develops reasonable drought resistance.
Feeding Light application of fowl manure or native plant food is acceptable in spring.
Pruning Branches may be reduced to about one-third after flowering.
Propagation Seed or cuttings.
Possible Problems Usually trouble free.

Dendrobium

SEE PAGE 222

Dianella caerulea

Paroo Flax Lily
FAMILY LILIACEAE

Perennial herb forming loose spreading clumps with glossy green tapering strap-like leaves 15–50 cm long, folded lengthways and fused towards the base in fan formation, either at ground level or more usually aerially. Erect slender wiry stems carry a long loose panicle of dainty pendulous blue flowers, 1–1.5 cm across, during spring or summer followed by glossy green berries which mature to bright blue.

Dianella caerulea

Origin All States except SA.
Size 0.3–1 m tall, from 0.3–2 m wide.
Soil Composted sand to moderately heavy loam with reasonable drainage.
Climate Cool to tropical; good salt wind and frost resistance.
Aspect Prefers lightly broken sunlight but also flowers in moderately dense shade to full sunlight.
Water Abundant to moderate; good drought resistance.
Feeding Complete or slow release fertiliser should speed growth.
Pruning Not necessary except for removal of old growths for thinning.
Propagation Seed or division of clumps.
Possible Problems Rarely affected.

Dianella tasmanica

Dianella laevis

Smooth Flax Lily
FAMILY LILIACEAE

Perennial forming dense clumps of smooth light to grey-green strap-like leaves, 20–80 cm long and to 1.5 cm wide, the wiry flower stems branching into a fine filament suspending pale violet flowers 1–2 cm across with prominent yellow anthers; flowering is during spring and summer and this is often followed by aerial plantlets.

Origin All States (several variations in form).
Size 0.5–0.8 m high, to 1 m wide.
Soil From dry rocky sandstone or granite outcrops to moderately moist heavy loam.
Climate Cool to tropical.
Aspect Sun to light shade.
Water Liberal but once established has good drought resistance.
Feeding Slow release or organic food should speed growth.
Pruning Rarely necessary, old flower stems may be removed.
Propagation Seed, division or carefully potted aerial plantlets.
Possible Problems Usually trouble free.

Dianella tasmanica

Tasman Flax Lily, Blueberry
FAMILY LILIACEAE

A handsome perennial for shaded areas, forming dense flax like clumps of broad strap-shaped leaves 0.5–1 m long and to 4 cm wide with a prominently keeled midrib. Flowers blue, to 1.5 cm across, in loose panicles, followed by glossy bright violet or blue berries 1–2 cm long.

Origin Tas., Vic. and NSW.
Size 0.5–1.75 m high, 0.5–2 m wide.
Soil Prefers moderately moist loam but adaptable.
Climate Cool to warmer temperate.
Aspect Full to dappled shade.
Water Prefers abundant but possesses excellent drought resistance.
Feeding Slow release boosts foliage.
Pruning Remove old or damaged leaves.
Propagation Seed or division of rhizomes.
Possible Problems Not usually affected but leaves become burnt and tattered when exposed to heavy frost with full sunlight.

Dillwynia prostrata

FAMILY PAPILIONACEAE

An attractive rock garden subject with low spreading or contour following branchlets well covered with glossy dark green slender leaves only about 0.5 cm long and a liberal scattering of deep golden yellow and red pea flowers about 1 cm across.

Dianella laevis

Dillwynia prostrata

Origin Mainly the higher country of north-eastern Vic. and south-eastern NSW.
Size 3–4 cm high, 1–1.5 m wide.
Soil Most well-drained light to medium loams.
Climate Cool to temperate. Good frost tolerance.
Aspect With either lightly dappled shade or morning sun.
Water Moderate at least until established.
Feeding Sparingly with organic or slow release fertiliser in spring.
Pruning Rarely required.
Propagation Layering, cuttings or scarified seed.
Possible Problems Usually trouble free.

Dillwynia pungens

WA Parrot Pea

FAMILY PAPILIONACEAE

Small erect shrub with very slender rigid and Prickly leaves to 2 cm long. The pea shaped flowers about 1 cm across have rounded yellow standards with reddish keel and wings and are clustered in leafy racemes from the leaf axils towards the ends of branches during mid-springtime early summer.

Origin Southern WA.
Size 1–1.5 m high about 1 m wide.
Soil Gravelly sand to well drained light loam.
Climate Cooler temperate to semi-tropical.
Aspect Broken sunlight to full sun.
Water Moderate.
Feeding Fowl manure or blood and bone sparingly in spring.
Pruning May be pruned back to about one-third after flowering.
Propagation Seed or cuttings.
Possible Problems Usually trouble free.

Dillwynia pungens

Dillwynia retorta

(syn. *Dillwynia ericifolia*)

Eggs and Bacon

FAMILY PAPILIONACEAE

Like many other shrubs with yellow and red pea flowers, it is commonly known as eggs and bacon. It is a delightful shrub with soft heath-like slightly twisted deep green leaves about 1 cm long slightly veiling the slender branches. Clusters of soft golden yellow pea flowers, about 1.5 cm across, blotched red between the standard petals, are carried terminally and in the leaf axils and brighten the bushland during late winter and spring.

Origin Qld, NSW and Vic., particularly in coastal sandstone areas.
Size Usually less than 1 m occasionally to 2 m and to 1 m wide.
Soil Well drained, preferably sandy with leafmould mulch.
Climate Cool to semi-tropical.
Aspect Best in broken sunlight or light shade.
Water Moderate.
Feeding Light dressing of blood and bone or other organic plant food in spring.
Pruning Prune back to one-half to one-third after flowering.
Propagation Cuttings of new growth as plant is losing sappiness.
Possible Problems Roots easily damaged by overfeeding, disturbing cultivation or poor drainage.

Dodonaea viscosa

(syn. *Dodonaea angustissima, Dodonaea attenuata*)

Hop Bush

A species widely varying from forms with small club-shaped leaves less than 2 cm long through slender linear to elliptic 17 cm long. The form illustrated, with lance-shaped to narrow oblong stemless (sessile) leaves to 10 cm long, was photographed in Mt Olga Gorge, NT. Like all forms the inconspicuous flowers are without petals and are followed by clusters of showy capsules often with leathery wings which turn reddish bronze where exposed to light; flowers mainly from late winter to summer.

Origin This form may occur in all States.
Size 2–4 m high, 2–3 m wide.
Soil Usually in rocky or gravelly areas but adapt to most well-drained loams.
Climate All areas from seacoast to arid inland.
Aspect Best with at least half sunlight. Good frost and wind tolerance.

Dillwynia retorta

Dodonaea viscosa

Water Moderate until established then very drought resistant.
Feeding Light applications of complete or slow release fertilisers will speed growth.
Pruning Prune back by one-third to one-half for more compact growth.
Propagation Seed or cuttings.
Possible Problems Scale, indicated by presence of black sooty mould.

Doryanthes excelsa

Gymea Lily, Gigantic Lily, Torch Lily

FAMILY AGAVACEAE

A dramatic plant making a large but graceful rosette of glossy bright green leaves 2–3 m long and to 20 cm wide. From the rosette grows a giant erect flower spike 4–7 m high. The bud, which is sheathed in red, resembles the classical flaming torch; it opens to a large branched cluster, 30–35 cm across, of six-petalled waxy flowers each 12–14 cm across, deep red on the outside, rosy above; flowers mainly in spring and summer but infrequently in cultivation, perhaps because bushfires are absent.

Origin NSW, mainly on sandstone areas of the Central Coast and Illawarra area.
Size 2–2.5 m high and similar width.
Soil Any fairly well-drained sandstone or loamy soil.
Climate Temperate to semi-tropical; heavy frost may damage flower buds.
Aspect Full to broken sunlight under tall trees.
Water Ample to moderate; good drought resistance.
Feeding Leafy mulch seems sufficient but organic or slow release fertilisers in spring should benefit.
Pruning Old flower spikes may be removed.
Propagation Seed, or division of clumps, which is difficult because root crop may be 40–50 cm below surface.
Possible Problems Usually trouble free.

Doryanthes palmeri

FAMILY AGAVACEAE

A spectacular plant similar to *D. excelsa* but foliage slightly broader, bolder and more grooved. Flower spike to 2 m long, arching rather than erect, with small clusters of erect reddish brown flowers 6–10 cm long on its upper side, mainly in summer.

Origin Mainly Qld and northern NSW.

Doryanthes palmeri

Size 2–3 m high and nearly as wide.
Soil Any moderately drained loamy soil.
Climate Temperate to semi-tropical; grows well in Melbourne Botanic Gardens in full sun.
Aspect Full to broken sunlight.
Water, feeding, pruning, propagation and possible problems, same as *D. excelsa*.

Doryanthes excelsa

Dryandra cirsioides

Dryandra cuneata

Dryandra cirsioides

FAMILY PROTEACEAE

Small shrub with hairy branches, slender sharply toothed leaves to 12 cm long, dark green above, whitish beneath; showy bright yellow flowers to 5 cm across, on short branchlets often closely arranged towards the end of a main stem; appearing from mid-winter to mid spring.

Origin WA, along the southern and western coast as far north as Geraldton.

Size 1–2 m high, to 1.5 m wide.
Soil Gravelly sand or light loam with good drainage.
Climate Cooler temperate to semi-tropical. Good frost resistance.
Aspect Full to lightly broken sunlight; not too crowded.
Water Moderate; good drought resistance when established.
Feeding Sparing; see *D. polycephala*.
Pruning Cut back after flowering.
Propagation Cuttings but seed preferable.
Possible Problems Seems fairly trouble free.

Dryandra formosa

Dryandra cuneata

Wedge-Leafed Dryandra

FAMILY PROTEACEAE

A compact dense foliaged and eye-catching dryandra with stiff dark green wedge-shaped leaves from 4–8 cm long deeply serrated or saw toothed and with wavy margins in some ways suggesting holly. The bright greenish yellow flower heads, to 5 cm across, with their long tapering stamens, resemble coarse textured pincushions.

Origin Southern coast of WA.
Size 1–3 m, high 1–2 m wide
Soil Naturally found in deep sand or gravelly soil but adapts to all but heavy loam if well drained.
Climate Temperate to cooler temperate except where frosts are very severe.
Aspect Accepts light shade but growth is more attractively compact when grown in at least two-thirds sunlight.
Water Moderately; once established it develops good drought resistance.
Feeding Use organic or native plant food. Avoid basic complete fertilisers with high phosphorus content.
Pruning Tip prune young plants for bushier growth; once established may be heavily pruned after flowering.
Propagation Seed.
Possible Problems Usually trouble free if soil does not remain soggy for long periods. Avoid heavy surface mulches in slow drying soil.

Dryandra formosa

Showy Dryandra

FAMILY PROTEACEAE

An erect shrub, with decorative long slender leaves evenly divided to the midrib in sawtooth fashion, deep grey-green above, downy underneath. Showy orange yellow flowers to 10 cm across are displayed in mid-spring. A popular cut flower, and more adaptable than some dryandra species from dry summer areas.

Origin WA, in the Stirling Range area.
Size 3–5 m high, to 3 m wide.
Soil From areas with peaty sand but adapts to most moderately drained light loams.
Climate Cooler temperate to semi-tropical with fair frost tolerance.
Aspect Broken to full sunlight; stands closer planting then other dryandras.
Water Moderately; give occasional soakings during long dry periods.
Feeding See *D. polycephala*.
Pruning May be hard pruned as flowers finish but without completely denuding stems of foliage.
Propagation Seed or cuttings.
Possible Problems As *D. polycephala*.

Dryandra polycephala

Many Headed Dryandra

FAMILY PROTEACEAE

A fairly large and attractive shrubby dryandra that adapts well to cultivation. The slender leaves to 20 cm. long are deeply and widely cut leaving sharply pointed triangular teeth widely spaced along a narrow midrib section. The deep yellow flowers, which remain colourful for several months, are about 4 cm across and occur on short branchlets along the erect stems; flowers in late winter to late spring.

Origin WA, from this Darling Range to about 300 km north.
Size 2-3 m high, to 2 m wide.
Soil Well-drained, not too acid, gravelly sand or light loam.
Climate Cooler temperate to semi-tropical, except in wetter areas of humid east coast unless grafted on to *Banksia integrifolia* or *B. spinulosa* which has been done with apparent success.
Aspect Broken to full sunlight, not too crowded by other plants.
Water Moderate; accustomed to fairly dry summers and withstands drought.
Feeding Sensitive to most complete fertilisers high in phosphorus. A leafy mulch and perhaps a scattering of compost or cow manure in spring is adequate.
Pruning Tip prune young plants; may be cut back below flowers when they finish.
Propagation Seed or cuttings.
Possible Problems Over fertilising or bad drainage are most likely causes of trouble.

Dryandra speciosa

Dryandra polycephala

Dryandra speciosa

Shaggy Dryandra

FAMILY PROTEACEAE

An oddity amongst dwarf shrubs with its low branches, its erect wiry leaves like pine needles and its almost animal-like, bell-shaped pendant flowers 4-5 cm long, covered in dull grey and brown, hairy to slender feather-like bracts hiding the soft reddish orange flowers. These appear from early winter to mid-spring.

Origin Dry summer gravelly areas of WA.
Size 0.5-1.5 m and as wide.
Soil Well-drained gravelly sand or gritty light loam.
Climate Temperate to drier semi-tropical. Resents east coast summer humidity.
Aspect Broken to full sunlight with good air circulation. May be more successful as a container plant.
Water Moderately; allow surface soil to dry between waterings.
Feeding Light application of slow release or organic fertiliser.
Pruning May be lightly pruned after flowering.
Propagation Seed or cuttings.
Possible Problems Subject to root fungus during prolonged summer wet periods.

Drypetes lasiogyna

FAMILY EUPHORBIACEAE

A small densely foliaged, rounded to umbrella-shaped tree with greyish brown bark that flakes to leave a pleasant mottling; bright green glossy elliptical leaves to 10 cm long, resemble bay leaves but are brighter. Juvenile leaves are toothed, with some resemblance to holly. The yellow flowers are about matchhead size but

Drypetes lasiogyna

female trees produce decorative glossy fruits resembling elongated cherries on short erect stalks; they change from green through orange to bright tomato red.

Origin Tropical to semi-tropical Australia, including Lord Howe Island and New Guinea.
Size 5–6 m high, 4–5 m wide, but much taller in moist rainforest areas.
Soil Light to medium well-drained loam.
Climate Temperate to tropical except very dry inland.
Aspect Needs the light shade from other plants initially then accepts full sunlight.
Water Ample to moderate.
Feeding Soluble fertiliser in spring and summer speeds early growth, then fowl manure or complete plant food for prolific fruiting.
Pruning Not required but trim and tip prune if preferred.
Propagation Seed, which must be sown fresh. Air layering is probably successful and if so would be preferable to assure sex of trees.
Possible Problems Scale, denoted by black sooty mould, is possible but otherwise seems trouble free.

Note: *D. subcubica*, a smaller growing species from Qld has 2 cm long fruit, almost cube shaped with four grooves running down either side.

Elaeocarpus reticulatus

Blue Berry Ash

FAMILY ELAEOCARPACEA

A quick growing and usually erect slender pyramid-shaped evergreen. 'Pagoda' shape describes it best because growth usually comes in tiers that gradually diminish as it ascends: Dark green, slender oblong to lance-shaped leaves to 11 cm long with slightly serrated margins; flowers like fringed white lily-of-the-valley occur in profuse sprays, mainly from older branchlets, in late spring. These are followed by deep blue 'berries' nearly 1 cm across which hold for a long time. There is a form with pinkish flowers and also *E. obovatus*, a larger growing species with prominently toothed shorter obovate leaves and slightly larger berries; it grows along the Qld and NSW coast.

Origin Semi-tropical and temperate east coast of Australia.
Size 6–12 m tall 4–5 m wide at head height then tapering. *E. obovatus* to 20 m high.
Soil Occurs mainly in leaf-mulched sandstone but adapts to most well-drained loams.
Climate Sub-tropical and temperate

Elaeocarpus reticulatis (berry)

Elaeocarpus reticulatus

coast, also inland with up to moderate frosts if not too dry.
Aspect Part shade, dappled or full sunlight.
Water Growth more rapid with ample water; reasonable drought resistance once established.
Feeding Growth may be speeded with monthly soluble plant food applications in spring and summer or slow release fertiliser.
Pruning Usually not needed but may be cut to hedge or other shape.
Propagation Berries germinate best when passed through birds; otherwise place ripened fruit under moist rotting leafmould. Cuttings may also be successful.
Possible Problems Rarely affected but thrips may silver leaves in sheltered positions during very dry conditions.

Embothrium

See *Oreocallis*

Epacris breviflora

Drumstick Heath

FAMILY EPACRIDACEA

An erect clump of rigid, vertical stems clad in sharply tapering elliptical leaves to 1.5 cm long, the upper sections liberally studded with sweetly fragrant white to blush pink starry flowers about 1 cm across; unlike most other species, their tubular base is not pronounced; flowers mainly in spring and early summer.

Origin Eastern Australia, especially in moist mountain areas.
Size 1–2 m high and nearly as wide.
Soil Sand or light loam with compost or moistened peatmoss added, moist but well drained.
Climate Cold to temperate.
Aspect Light shade or broken sunlight but accepts full sun in cool districts.
Water Ample but let surface dry occasionally between waterings.
Feeding Sparingly if at all with slow release or half strength water-soluble plant food.
Pruning May be pruned back to about half as main flowering flush finishes; cut out old branches.
Propagation Cuttings or seed sown in moist mossy ground.
Possible Problems Usually trouble free.

Epacris impressa

FAMILY EPAERIDACEAE

The floral emblem of Victoria. One of the most eye-catching of the larger flowered epacris. Small semi-erect shrub with 1–1.5 cm, stem clasping oval to lance-shaped shiny green leaves tapering to a sharp point. A profusion of waxy textured tubular flowers 2 cm long are pendant from leaf axils towards the ends of branches. These are frequently rosy red but there are pinks, whites and combinations of these colours; flowers mainly during winter.

Origin SA, Tas., Vic. and NSW.
Size 0.5–2 m high, to 1 m wide, usually more compact in sun.
Soil Prefer fairly moist but reasonably drained peaty loam.
Climate Temperate to cool.

Epacris breviflora

Epacris impressa

Aspect Broken sunlight, light shade or up to two-thirds sun. Prefer protection of other plants and a good leafy mulch.
Water Ample is ideal but once established withstands periods of dryness.
Feeding Avoid strong fertilisers. Safest to mulch in spring with leafy compost, previously heaped cow manure. Seaweed extract or half-strength complete soluble plant food at monthly intervals in spring is also acceptable.

Pruning: After flowering prune back to leave only a few centimetres of leafy growth on stems then, if desired, tip prune later to encourage still bushier growth.
Propagation Cuttings, transplanting with care when rooted.
Possible Problems Usually trouble free except for leaf tip-webbing caterpillar which merely has tip-pruning effect.

Epacris longiflora

Epacris microphylla

Coral Heath

FAMILY EPACRIDACEA

Small erect plant with its wiry branches densely clad in tiny stem-clasping and prickly sharply tapering leaves to 5 mm long. Small white or pink tinged bells slightly shorter than the leaves fill the leaf axils along the upper section of stems from autumn to mid-spring.

Origin All eastern Australian States, especially along mountain and coastal districts.
Size 0.5–1.5 m high rarely more than 60 cm wide.
Soil naturally occurring both in sedgy and moderately dry light sandy or medium loam with leafy mulch, sometimes only in a layer of moist moss over rock.

Epacris microphylla

Epacris longiflora

Native Fuchsia,
Fuchsia Heath

FAMILY EPACRIDACEA

A delightful epacris with lateral to pendulous branches unless supported by surrounding shrubbery or pruned. Similar to *E. impressa* but the glossy leaves are more heart shaped and abruptly tapering, flowers slightly longer and the tube changes from rosy red to cream or white as it nears the flared tip petals. A beautiful container plant.

Origin NSW, mainly in coastal sandstone, sometimes clinging to damp mossy rock faces or in lightly timbered bushland. A form with semi-erect flowers occurs in southern Qld.
Size 0.3–1.5 m high, to 2 m wide (see description).
Soil Moist but well-drained sand or loam with a leafy mulch.
Climate Cool to semi-tropical; has good frost resistance.
Aspect Best with at least light sun protection but accepts full sun or shade.
Water Prefers moist soil but stands periods of dryness.

Feeding, pruning, propagation and possible problems as for *Epacris impressa*.

Eremaea beaufortioides

Aspect, water, feeding, pruning, propagation and possible problems as for *E. impressa*.

Eremaea beaufortioides

Round Leafed Eremaea

FAMILY MYRTACEAE

A fairly dense and attractive shrub, the stems covered with smooth, mostly rounded, grey-green leaves to 6 mm long in crowded overlapping rows. Flowers like small bright reddish orange gum blossoms with golden anthers, 1–1.5 cm across, are clustered in showy terminal heads of three to five.

Origin Mainly the sandy plains from the Darling Range to about 200 km north.
Size 1–3 m high and almost as wide.
Soil Light to medium, not too acid, well-drained loam, if necessary made crumbly by the addition of coarse sand and leafy surface mulch.
Aspect Prefers full sunlight and an open position with good air circulation.
Water Moderate until established then withstands dryness.
Feeding Light dressing of blood and bone and fowl manure in spring.
Pruning Prune back to about two-thirds after flowering.
Propagation Seed or from just firmed cuttings of new growth.
Possible problems Subject to fungus disease in moist humid areas, especially if crowded.

Eremophila christophori

FAMILY MYOPORACEAE

A many branched shrub tensely clad in soft green lance-shaped to elliptical leaves 3 cm long, and numerous 2 cm long tubular blue flowers from leaf axils at the end of branchlets mainly from late autumn through until spring.

Origin Central Australia, mostly MacDonnell Range area.
Size 1–1.75 m, to 2 m wide.
Soil Well-drained and slightly limey sandy to gravelly soils with a leafy mulch.
Climate Temperate to drier tropical; good frost resistance.
Aspect At least half sunlight.
Water Moderately until established then good brought resistance.
Feeding sparingly with slow release or very light scatterings of animal manure.
Pruning Branches may be shortened back by about one-third as flowering subsides. For more compact growth, tip prune when young; tip prune new growth after major pruning.
Propagation Cuttings or from seed which may be slow to germinate.
Possible problems usually trouble free in dry climates.
Family Myoporaceae

Eremophila christophori

Eremophila divaricata

Eremophila divaricata

Silver Emu Bush

FAMILY MYOPORACEAE

The silvery foliage of this small shrub makes it an excellent foil or contrast for other plants. The dense intertwining branches are a powdery silvery-grey likewise the 1–1.5 cm slender leaves. Short bell-shaped lavender blue to pale

Eremophila gilesii

Eremophila glabra

violet flowers are clustered towards the tips of branches from spring to mid-summer.

Origin SA, Vic. and NSW.
Size 1-2 m high and as wide.
Soil A wide range of heavy or light loams with reasonable drainage.
Climate Cool temperate to semi-tropical. Grows in the wet summer areas as well as more arid regions.
Aspect Full sun to broken sunlight.
Water Moderate; accepts prolonged dryness when established.
Feeding Slow release or complete plant food sparingly in spring or late summer.
Pruning May be pruned heavily as flowers finish.
Propagation Seed or cuttings.
Possible problems usually trouble free.

Eremophila gilesii

Desert Fuchsia

FAMILY MYOPORACEAE

A small fairly open shrub with wiry yellowish brown branches and stiff slender foliage to about 8 cm long, usually deeply channelled above and sticky when young. The short tubular flowers extend 2-2.5 cm across with broadly ovate and fluted mauve or purple petals. These appear mainly from early winter to late spring.

Origin More arid areas of NT, WA, SA, western Qld and NSW.
Size 1-1.5 m high and similar width.
Soil Well-drained slightly limey and lightly composted sand or gritty loam.
Climate Temperate to semi-tropical; good frost resistance; more suited to arid or semi-arid regions.
Aspect Open sunny position.
Water Moderately until established; excellent drought resistance.
Feeding Sparingly with slow release, complete native plant food or fowl manure.
Pruning May be pruned back heavily after flowering.
Propagation Seed or cuttings.
Possible Problems Usually trouble free; regarded as a pest of pasture in some areas especially south-western Qld because of its stubborn regeneration after slashing.

Eremophila glabra

Emu Bush, Tar Bush

FAMILY MYOPORACEAE

Probably the most widely grown eremophila and most variable in foliage and flower colour. They are mostly small shrubs with dense spreading growth; stems whitish; alternate leaves to 5 cm long are linear to lance shaped, sometimes broadening almost to elliptical, usually smooth with entire margins as the type illustrated, but can be hairy and lightly toothed. The tubular flowers are 3-4 cm long with stigma and stamens prominently protruding (exserted) and colour forms range from green to yellow, orange or red, appearing mainly from late winter to late summer.

Origin All States excepting Tas., mainly in semi-arid to arid areas.
Size 1-1.75 m high, 1.5-3 m wide.
Soil Most well-drained light soils, preferably slightly limey.
Climate More suitable for drier temperate to semi-tropical areas but adapts to all but very wet and humid or extremely frosty areas.
Aspect At least half sun.
Water Moderately until established; good drought resistance.
Feeding Responds to slow release or complete plant foods in moderation.
Pruning May be pruned heavily after flowering.
Propagation Seed or cuttings.
Possible Problems Flower mould in wet summer climates which can extend to and kill branches. Spraying with mancozeb gives some control.

Eremophila maculata

Native Fuchsia,
Spotted Emu Bush

FAMILY MYOPORACEAE

An attractive bushy and usually fairly upright shrub with smooth lance-shaped to slender elliptical leaves to 5 cm long and prominent tubular flowers 2.5-4 cm long with upper petals extended like a

Eremophila maculata

Eremophila ovata

hood and lower rolled under to expose the spotted throat and extended stamens. Leaf colour may vary according to variety from deep green to grey-green and flowers from white through yellow, orange-red to purple. Flowering time is mainly from winter to late spring.

Origin The semi-arid to arid areas of all mainland States.
Size 0.75–3 m high and as wide.
Soil Preference for slightly limey heavier loams but also succeeds in well-drained lighter soils.
Climate Temperate and semi-tropical; good frost resistance.
Aspect Best with at least half sun; ideal for exposed windy positions.
Water Accepts occasional flooding but has excellent drought resistance when established.
Feeding Accepts light applications of complete plant foods or fowl manure.
Pruning May be pruned fairly heavily after flowering. Tip prune when young to encourage compact spreading growth.
Propagating Seed or cuttings of maturing younger growth.
Possible Problems Few pest problems except mould may attack flowers and stems during humid conditions. Young growth has high cyanide content and may be toxic to stock, especially after rain.

Eremophila maculata var. *brevifolia*

Eremophila maculata var. *brevifolia*

FAMILY MYOPORACEAE

Different in appearance to the species described as growth is usually denser with small, grey, broad oval (orbicular) leaves rarely more than 1 cm long and the flowers bright ruby red.

Origin NT into WA and possibly western Qld.
Size 0.5–1.5 m high, to 2 m wide.
Soil Grows well on both slightly alkaline gritty or gravelly and clayey loams.
Climate Drier temperate to semi-tropical districts.
Aspect Best in full sunlight with good air circulation.
Water Moderately until established; good drought resistance.
Feeding Complete plant food, blood and bone or fowl manure applied sparingly.
Pruning For more compact growth prune back by one-third to one-half after flowering.
Propagation Seed or maturing new growth.
Possible Problems As the species but fungus infection of flowers and growth is even more likely during humid or wet spring to early summer conditions.

Eremophila ovata

FAMILY MYOPORACEAE

A low spreading semi-prostrate shrub with slender hairy yellowish buff branchlets, grey-green oval to lance-shaped leaves, 2–3 cm long and short but broad tubular lilac or sometimes scarlet flowers to 2.5 cm long during mid winter to late spring.

Origin NT.
Size 0.5–0.75 m high, to 1 m wide.
Soil Well-drained slightly alkaline gritty loam.
Climate Drier tropical to temperate areas.
Aspect Full to lightly broken sunlight.
Water Moderately until established then sparingly.
Feeding Light applications of complete plant food or animal manure in spring.
Pruning Prune lightly after flowering.
Propagation Seed or cuttings of maturing new growth.
Possible Problems Subject to root rot in prolonged wet or humid conditions. Even the plant illustrated, growing in the Olive Pink Reserve, Alice Springs, had been grafted on to myoporum stock to give greater vigour and resistance to root fungus.

Eriostemon buxifolius

Eriostemon australasius

Eriostemon australasius

(syn. *Eriostemon lanceolata*)
Pink Wax Flower
FAMILY RUTACEAE

A beautiful eriostemon; if not too crowded it has an attractively compact dome shape, with slender oblong to lance-shaped waxy light to mid-green foliage to 7 cm long, liberally spangled with large but beautifully simple waxy pink five-petalled flowers about 4 cm across, mainly in upper leaf axils of outer branchlets. Colour varies from pale mauve-pink to soft rose. Flowers from mid-winter into early spring. Long lasting cut flower.

Origin Southern Qld and NSW, especially on sandstone hillsides.
Size 1–2 m high to 1.5 m wide.
Soil Adapts to heavier loams if lightened by leafmould as well as sandstone regions if well drained.
Climate Semi-tropical to cooler temperate.
Aspect Lightly broken to near full sunlight. Protect from root disturbance.
Water Moderate to ample but accepts periods of dryness.
Feeding Light dressings of slow release fertiliser or fowl manure in spring.
Pruning Seems best if stems are pruned back to just above the few lowest leaves when flowers finish.
Propagation Not easy! Seed abraided between sandpaper and well washed has given some results. Cuttings are slow and uncertain. Layering is worth trying.
Possible Problems Apart from an occasional leaf-webbing caterpillar, usually trouble free.

Eriostemon myoporoides

:riostemon difformis

Eriostemon buxifolius

3ox Leaf Eriostemon

'AMILY RUTACEAE

\ small many branched shrub, very ippealing when in flower, with closely et stemless (sessile-elliptical leaves ibout 1 cm long, resembling those of)uxus (box hedge); outer branches :lustered with comparatively slender)etalled flowers to about 2.5 cm across, leep pink in bud opening to white lushed pink during late winter and spring. An excellent container plant. There is a double flowered form known is 'Parry's Pink'.

)rigin Temperate NSW coast on andstone ridges.
ize 1–1.3 m high, 1–2 m wide.
Soil Native to sandstone but adapts to well-drained light loam.
Climate Temperate; good frost resistance.
Aspect Lightly broken or to about half direct sunlight. Good wind tolerance.
Water Prefers moderate watering, especially in late spring and summer.
Feeding Slow release or organic fertiliser in spring.
Pruning Shorten back branches after flowering and for more compact growth, tip prune new shoots.
Propagation Cuttings, which are slow to make root.
Possible Problems Rarely affected.

Eriostemon difformis

FAMILY RUTACEAE

A spreading but usually fairly compact shrub with slightly warted and hairy branches and small slender fleshy leaves to about 1 cm long usually with rolled margins. The bush is massed with terminal clusters of stemless, little white starry flowers about 1 cm across, mainly during autumn and winter. There are several forms: sub sp. *smithianus* with broader foliage may be found in cultivation and reputedly the most adaptable.

Origin Qld, NSW and Vic, mainly in slopes of Dividing Range.
Size 1–2 m high and similar width.
Soil Well-drained sandy or light loam with a leafy mulch.
Climate Temperate to cooler temperate, good frost tolerance.
Aspect Full to lightly broken sunlight.
Water Moderate; the species appears to resent prolonged wet during summer. Tolerates dry periods well.
Feeding Slow release fertiliser or fowl manure sparingly in spring and autumn.
Pruning May be pruned back lightly after flowering.
Propagation Cuttings which are slow to make root.
Possible Problems Apparently the species especially is subject to root fungus in many areas; sub sp. *smithianus* evidently has greater resistance.

Eriostemon myoporoides

Long Leafed Wax Flower

FAMILY RUTACEAE

A well-known garden shrub, popular for its adaptability, compact form and masses of waxy textured starry flowers, mainly from late winter to mid-spring. The widely accepted form has slender oblong to lance-shaped deep blue-green leaves to about 11 cm long with clusters of pink tinted buds, opening to starry white flowers 1.5 cm across, but there are forms with elliptical and short wedge-shaped to rounded foliage.

Origin Vic., NSW and Qld, particularly in the lightly treed foothills of Dividing Range.
Size 1–1.5 m high and as wide.
Soil Adapts to most well-drained and preferably mulched soils.
Climate Cool to semi-tropical; good frost resistance.
Aspect Light shade to full sunlight.
Water Moderate; once established accepts dry periods.

Eriostemon obovalis

Feeding Accepts most complete fertilisers in moderation.
Pruning May be pruned back to at least one-third after flowering. Tip prune for compact form.
Propagation Cuttings
Possible Problems In some areas is occasionally attacked by scab, causing black sooty mould on stems and foliage.

Eriostemon obovalis

FAMILY RUTACEAE

An open shrub with slender, arching, slightly warted stems and obcordate (heart shape reversed) leaves, 0.5–1 cm long, warted underneath. Flower buds in the leaf axils, lightly flushed pink, opening to white, about 1.5 cm across, mainly during late winter and early spring.

Origin Blue Mountains, NSW.
Size 11.5 m high, to 0.5 m wide; a little more compact in full sun.

All other requirements as *E. buxifolia*

Note: *E. verrucosus* is very similar to *E. obovalis* but its leaves and flowers are a little larger. It originates from Vic. (especially the Grampians area), Tas. and SA. A beautiful double form, 'Mrs J. Semmen' is sometimes available. Growth is relatively pendulous.

Eriostemon obovalis 'Mrs J. Semmen'

Eucalyptus

Eucalyptus Reclassification: Although this genus has been undergoing reclassification, all included in this text appear under Eucalyptus, as this will undoubtedly remain their horticultural title for many years. For readers interested, the present new botanical generic name is bracketed after each entry under Eucalyptus in the index.

Eucalyptus burdettiana

(Symphyomytus burdettiana)

Burdett Mallee

FAMILY MYRTACEAE

A quick growing interesting and attractive small tree or large mallee-type shrub with dense, dark green, leathery, lance-shaped foliage 8–12 cm long and finishing with a long thin tip. The slender greenish yellow finger like buds are about 5 cm long, in clusters of three to seven, opening to broad, pendulous, greenish cream flowers to 4 cm across, and followed by 2.5 cm long, bell-shaped brown capsules with a prominent rib. The smooth bark is bronze-green to light brown, and sheds itself in late summer.

Origin South-eastern WA.
Size 4–9 m high with a spread of 4–5 m, or occasionally wider.
Climate Best suited to more arid temperate to semi-tropical regions.
Soil Adapts to a wide range of well-

Eucalyptus burdettiana

Eucalyptus caesia

drained slightly acid to alkaline soils.
Aspect Best with at least two-thirds sun.
Water Moderate until established.
Feeding Slow release or complete plant food applied sparingly when established.
Pruning Not necessary but can be cut hard back for coppice-type growth.
Propagation Seed.
Possible Problems Usually trouble free in well-drained situations.

Eucalyptus caesia

(*Symphyomytus caesia*)

Gungurru

FAMILY MYRTACEAE

A delightful small tree with silver grey powdery branchlets and clusters of pendant large red winter-spring flowers with silver calyces and stems followed by silver fruits (gum nuts). Also the slatey bronze bark of the trunk splits and curls to reveal contrasting greenish new bark. Sub sp. *magna* has larger flowers and greyer and more pendulous growth.

Origin Granite outcrops of southern WA wheat belt.
Size 4-8 m high, 3-6 m wide; sub sp. *magna* to nearly twice this height.
Soil Adapts to light and clayey loams providing they are well drained, lightened with compost and not excessively acid.
Climate Temperate; moderate frost resistance.
Aspect Full sunlight but if possible some wind protection.
Water Moderately until established; good drought tolerance.
Feeding Restrained use of organic or slow release fertiliser in spring–summer providing conditions are not excessively dry.
Pruning Pruning to thin foliage during early growth and removal of excess seed capsules minimises risk of splitting and other wind damage.
Propagation Seed.
Possible Problems Scale pests. Weak and top heavy growth from excessive watering and feeding. If tree does blow over, leave to reshoot from lignotuber rather than attempt to prop up.

Eucalyptus calophylla

(*Corymbia calophylla*)

Marri

FAMILY MYRTACEAE

A dense, round headed tree closely resembling the popular *E. ficifolia* (WA Red Flowering Gum) but grows taller and leaves slightly narrower, to 3 cm across and 6-18 cm long, and ending in a fine tip. During late summer or autumn it displays a profusion of flowers to 4 cm across held in clusters slightly above the foliage, colour varying from white through cream to soft pink (variation from seed is inevitable). Flowers are followed by large decorative urn-shaped capsules to 5 cm long and 4 cm wide.

Origin Southern WA.
Size 12-25 m high, to 10 m high.
Soil Well-drained but not excessively sandy to clayey loam.
Climate Temperate to cooler temperate. Moderate tolerance to frost and seacoast conditions.
Aspect Full to part sun.
Water Appreciates ample water but not frequently soggy soils or long periods of dryness.
Feeding Complete plant food or slow release granules applied just beyond drip line of foliage will promote growth. However, continuous feeding and watering reduces root spread and can cause top heavy growth.
Pruning Only as needed for shape.
Propagation Seed which germinates readily.
Possible Problems Prone to leaf galls but without any serious effect.

Eucalyptus cephalocarpa

(*Symphyomytus cephalocarpa*)

(syn. *Eucalyptus cinerea*)

Argyle Apple

FAMILY MYRTACEAE

A small to medium ornamental tree grown mainly for its silver-grey foliage, especially the juvenile leaves which it usually retains for a number of years. These are opposite, stem clasping, rounded to heart shaped and blue-grey, 5-8 cm long. Adult leaves are broad lance shape, 7-11 cm long but still a bluish grey in colour. The invariably bent trunk is dark blackish brown, furrowed, with persistent bark. The

Eucalyptus calophylla

Eucalyptus cephalocarpa

white flowers, about 1 cm across, are tucked in the leaf axils during late spring to early summer.

Origin Vic. and mainly southern half of NSW in sandstone or granite, just west of Dividing Range.
Size 8–15 m high and in open situations nearly as wide.
Soil Adapts to most well-drained soils.
Climate Cool to semi-tropical.
Aspect Foliage sparse until it reaches at least half sunlight. Good wind resistance.
Water Moderate until established, then drought resistance is good.
Feeding Do not feed within three months of planting, then progress may be speeded by water soluble or slow release plant foods.
Pruning Optional, but hard pruning usually promotes more juvenile foliage.
Propagation Seed.
Possible Problems Occasionally disfigured by leaf rolling or leaf-skeletonising caterpillars.

Eucalyptus erythrocorys

(Eudesmia erythrocorys)

Illyarrie Gum, Red Cap Gum

FAMILY MYRTACEAE

Eucalyptus erythrocorys

A mallee-type gum grown mainly for its decorative bright yellow flowers and distinctive waxy red bud cups with their four lobes divided by a raised X. It may be a single trunked small tree with smooth grey-brown deciduous bark or make a clump of shorter growth. Young branches reddish, juvenile leaves, grey, opposite and oval to broad lance shape; mature leaves also generally opposite, leathery green, usually sickle shaped, to 25 cm long. The flowers, about 5 cm across, appear mainly during mid-summer and autumn. The large capsules are ornately ribbed.

Origin South and western coast of WA.
Size 4–10 m and may spread to 8 m.
Soil Most well-drained limey or slightly alkaline soils.
Climate Temperate to semi-tropical; young trees damaged by heavy frost.
Aspect Open sunny situation.
Water Moderate until established then drought resistant.
Feeding Sparingly in spring when established.
Pruning Single stemmed plants may be cut back hard to form coppice or clump effect.
Propagation Seed.
Possible Problems Foliage subject to attack by case moth and other caterpillars.

Eucalyptus ficifolia

(Corymbia ficifolia)

Red Flowering Gum

FAMILY MYRTACEAE

Undoubtedly the most popular ornamental eucalyptus, hopefully grown for its glorious summer canopy. 'Hopefully' because from seed, colour can range from white through pink to orange-red or deep scarlet. However, nurserymen normally select seed from trees with records of producing good colours. Young plants with red stems and leaf stalks usually produce the deeper colours. Juvenile leaves oval to broad lance shaped and alternate, adult leaves similar but with heavier texture, 7-14 cm long. Flowers to 4 cm across, followed by large urn-shaped seed capsules. Trunk normally solitary with rough scaley bark opening to a much branched dense rounded top.

Origin Southern WA, mostly on slightly acid gravelly sand over clay.
Size 5–9 m high with a crown spread of 4–7 m, sometimes larger in natural habitats.
Soil Most well-drained loams except salty or limey soils.
Climate Temperate; young plants damaged by heavy frost.
Aspect Need at least two-thirds sun to flower well; good wind tolerance.
Water Moderately until established. Best in winter rainfall areas but worth growing in eastern and southern coastal districts.
Feeding Once established is benefited by moderate feeding with animal manure, slow release or complete plant foods during late autumn or early winter.
Pruning Not necessary but branchlets may be thinned to allow more light penetration.
Propagation Seed.
Possible Problems Raised corky purple streaks are usually caused by tiny wasps but are not of consequence.

Eucalyptus haemastoma

(Monocalyptus haemastoma)

Scribbly Gum

FAMILY MYRTACEAE

A picturesque small tree, often developing two or three outward leaning trunks. These feature smooth whitish bark with large slaty grey patches and intriguing scribble-like markings made by the tunnelling of insect larvae below the previous season's bark. The broad lance-shaped leaves are 10–15 cm long; small white flowers appear mainly in summer but are spasmodic. A useful tree to provide the 'lightly broken' sunlight preferred by most boronias, dillwynias, etc.

Origin Sandstone ridges of NSW central coast and tablelands.
Size 6–12 m high; canopy spreading 7–10 m.
Soil Preferably light sandy or gravelly loam which is well drained.
Climate Cooler temperate to semi-tropical.
Aspect Best in fairly open sunny situations; good tolerance to salty winds.
Water Moderate; at least until established.
Feeding Naturally a low fertility tree but slow release or complete plant food applied sparingly will speed growth.
Pruning When well-established hard

Eucalyptus ficifolia

Eucalyptus haemastoma

cutting back should induce formation of other trunks.
Propagation Seed.
Possible Problems Scale may be a problem during early growth, rarely with maturity.

Eucalyptus macrocarpa

(Symphyomytus macrocarpa)

Rose of the West

FAMILY MYRTACEAE

A spectacular eucalyptus although only in the sprawling shrub category. Stems or 'trunklets' radiate laterally from the basal lignotuber then ascend, their new bark covered with powdery bloom and arrayed with four rows of blue-grey, opposite, broadly ovate to elliptical leaves to 12 cm long. Large oval flower buds the same glaucous colour come singly along the upper branches, lifting their caps to release huge flowers to 10 cm across, packed with either red or deep rose gold-tipped anthers. These are followed by button-shaped capsules about 9 cm across and about half as deep.

Origin WA, south-east of Geraldton down to south east of Perth.
Size 2–4 m high and to twice the width.
Soil Slightly acid or slightly alkaline deep crumbly light loam or deep sand over clay subsoil; good drainage essential.
Climate Temperate and semi-tropical but not the east coast. Best suited to semi-arid or dry summer areas.
Aspect At least two-thirds full sun; good frost tolerance.
Water Moderately until established then sparingly; withstands drought.
Feeding Sparingly; a light dressing of slow release food in autumn should speed growth.
Pruning Old woody stems may be removed at their base.
Propagation Seed germinates readily.
Possible Problems Caterpillars may attack foliage, overwet soil and waterlogging cause leaf blackening and ultimate loss of plant.

Eucalyptus maculosa

See *E. mannifera* sub sp. *maculosa*

Eucalyptus mannifera

(Symphyomytus mannifera)

sub sp. *maculosa*

Red Spotted Gum

FAMILY MYRTACEAE

Erect or low branching tree with attractively blotched deciduous smooth powdery bark, white to grey most of the year, reddish in early summer then cream. The slender grey-green foliage allows plenty of broken sunlight through and displays the graceful spreading branches. Initial growth rapid, then slow after 6–7 m. Cream flowers in summer. Widely planted in Canberra.

Origin Vic., northward along the tablelands to New England, NSW.
Size 10–15 m high, 5–15 m wide.
Soil Any well-drained soil. Occurs naturally on rocky granite to sandstone hillsides.
Climate Popular tree for street and park planting in tableland and other cool areas.
Aspect Open to lightly broken sunlight.
Water Moderate, at least until established; good drought resistance.
Feeding Slow release, complete fertiliser or fowl manure mainly in spring but only when well established and soil is moist.
Pruning Only if necessary for shape.
Propagation Seed.
Possible Problems Leaf-eating caterpillars and scale pests may occur during early growth, usually trouble free later on.

Eucalyptus sideroxylon

(Symphyomytus sideroxylon)

Mugga, Pink-Flowered Ironbark

FAMILY MYRTACEAE

A graceful and adaptable medium size to tall tree, the trunk with persistent dark furrowed tough bark, smooth outer branches gracefully pendulous with slender lance-shaped mature leaves to 14 cm long, either dull green or grey. Those of the cultivar sold as 'Rosea' are blue-grey, its pendulous flowers rosy pink, appearing during winter or spring. There are also white-, cream- and red-flowered forms which all seem to produce fairly reliably from seed. The White Ironbark, *Eucalyptus leucoxylon*, is similar to *E. sideroxylon* except that all but the base of its trunk is smooth brownish grey. The bark is shed in late summer with new bark creamy white. Leaves are usually greener and it is found in SA, Vic. and NSW.

Origin Vic., northward into south-eastern Qld; mostly west of coastal belt on shale or stony ground.
Size 10–20 m high, 6–10 m wide; sometimes larger in natural habitats.
Soil Most well-drained soils but prefers shaley or clay subsoil.
Climate Cool to semi-tropical.
Aspect Open, sunny or initially with broken sunlight.

Eucalyptus macrocarpa

Eucalyptus mannifera

Water Moderate but good drought resistance when established.
Feeding Slow release, complete plant food or animal manure when established, preferably in spring.
Pruning Only when necessary for shape.
Propagation Seed.
Possible Problems Leaf-eating caterpillars or scale pests sometimes occur.

Eucalyptus sideroxylon

Eucalyptus torquata

(Symphyomytus torquata)

Coolgardie Gum, Coral Gum

FAMILY MYRTACEAE

A delightful small tree, widely planted for the pleasure of its pendulous clusters of ornately ribbed reddish buds with beak-like caps and the 2.5–3.5 cm across pink to red flowers bursting from them. Branchlets are also a rich coppery red and are complemented by bluish grey lance-shaped leaves to 12 cm long. Another great advantage of this plant is that it usually commences flowering during its second or third year, mainly from early spring to mid-summer and sometimes much longer.

Origin Coolgardie district of WA.
Size It can grow 4–10 m high and 3–10 m wide but in cultivation the lower figure is more usual.
Soil At home on stony hillsides but accepts most well-drained mildly acid to alkaline soils.
Climate Ideal for dry inland temperate areas but also accepts cool temperate to sub-tropical. Life on humid east coast may not exceed ten years but is still worth trying. Reasonably frost hardy.
Aspect Best in open sunny areas; has good wind resistance.
Water Moderate until established; good drought resistance.
Feeding Sparingly with slow release, complete or animal manure during spring or autumn when soil is moist.
Pruning Responds to moderate pruning

Eucalyptus torquata

Eucalyptus leucoxylon (rosea) (see under *E. sideroxylon*)

Eucalyptus leucoxylon (white)

but rarely necessary.
Propagation Seed.
Possible Problems Scale may occur, particularly during early growth.

Eutaxia obovata

FAMILY PAPILIONACEAE

Small rounded shrub, the slender stems clad with closely set stiff dark green leaves to 2 cm long in four rows and during spring clustered with small yellow and red pea flowers.

Origin Southern WA.
Size 0.75–1 m high and nearly as wide.
Soil Adapts to a wide range of well-drained garden soils.
Climate Cooler temperate to tropical.
Aspect Full to dappled or half sunlight.
Water Moderate, at least until established and during early summer.
Feeding Light application of complete fertiliser in spring.
Pruning Prune back after flowering leaving at least a few sets of foliage at the base of stems carrying the old flowers. Tip prune new growth.
Propagation Scarified seed or cuttings.
Possible Problems Usually trouble free.

Everlasting Daisy

See *Helipterum*

Faradaya splendida

FAMILY VERBENACEAE

A spectacular and vigorous climber with rich green, glossy, oval to elliptical leaves to about 18 cm long and in spring displays clusters of fragrant white tubular flowers with protruding stamens and four broad petals or lobes spreading to 5 or 6 cm. Its vigorous and spreading habit makes it more suitable for large gardens.

Origin Qld rainforests.
Size Spread of 8–10 m with protruding canes.
Soil Adapts to most well-drained soils.
Climate Tropical to semi-tropical.
Aspect Shade or sun but flowers more freely with at least half sun.
Water Appreciates liberal water during summer.
Feeding If at all use a basic complete plant food sparingly as most others result in excessive growth.
Pruning May be pruned back after flowering and new growth restrained by tip pruning.
Propagation Cuttings or fresh seed.
Possible Problems Fairly trouble free in warm climates.

Flannel Flower

See *Actinotus*

Geebung

See *Persoonia*

Eutaxia obovata

Gompholobium grandiflorum

Golden Pea

FAMILY PAPILIONACEAE

A small bushy and attractive shrub with needle like leaflets about 2.5 cm long forming trifoliate leaves, sometimes bronze tinged and in spring displaying prominent pale golden pea flowers with a broad semicircular standard to 2.5 cm. in diameter; they either appearing singly or in terminal clusters of two or three. The buds or calyces are a contrasting blackish bronze.

Origin NSW, mainly in the Hawkesbury sandstone areas bounded by Newcastle, Wollongong and the Blue Mountains.
Size 1–1.5 m high and as wide.
Soil Well-drained sand or sandy loam with comport added and a leafy mulch.
Climate Cool to semi-tropical; tolerates moderate frost.
Aspect Prefers lightly broken to dappled sunlight.
Water Liberal to moderate; good drought resistance.
Feeding Sparingly with slow release or organic fertiliser when new spring growth follows pruning and soil is moist.
Pruning Keep more compact by lightly pruning after flowering then tip pruning new growth.
Propagation Cuttings or seed.
Possible Problems Leaf-rolling-caterpillar can damage buds or more usually the seed pods.

Gompholobium latifolium

Golden Pea,

Golden Glory Pea

FAMILY PAPILIONACEAE

This plant was obviously discovered and named after *Gompholobium grandiflorum* because it warrants the latter name, having the largest flowers of the genus. It has more erect growth, flatter and broader leaflets (about 2–6 mm wide) and to 2.5 cm long) slighter brighter flowers and larger, with a broad standard to 3 cm across. The nearly black buds and calyces are similar.

Origin Vic., NSW and Qld, from Dividing Range to coast, mostly in lightly timbered sandstone areas.
Size 1.5–2 m high and to 1 m wide, but more compact with pruning.
Soil Prefers well-composted sandy soils but adapts to all but heavy clayey loams providing drainage is good.
Climate Cooler temperate to semi-tropical. Tolerates all but very severe frost.
Aspect Best with lightly dappled sunlight or just a few hours morning sun.
Water Moderate to liberal but allow surface soil to dry between waterings; reasonable drought tolerance.
Feeding Sparingly if at all with slow release or organic fertiliser.
Pruning Prune well back after flowering but leave some foliage below most cuts, then tip prune new growth.

Faradaya splendida

Gompholobium grandiflorum

Gompholobium latifolium

Goodenia ovata

Hop Goodenia

FAMILY GOODENIACEAE

A soft wooded quick growing shrub often with irregular but mostly lance-shaped bright green leaves, 6–10 cm long, with toothed margins; young growth especially is glossy and usually sticky. Yellow flowers to 2 cm across are carried in axillary racemes from early spring into summer.

Origin All States except Tas. and WA.
Size 1–2 m high and spreading usually to about 1 m, sometimes twice as wide.
Soil Accepts all but very limey or alkaline soils; will tolerate even sedgy areas.
Climate Cool to semi-tropical.
Aspect Preference for likely broken sunlight or half sun.
Water Happiest with liberal water but withstands dry conditions.
Feeding If any, complete fertiliser used sparingly; most nitrogen-rich mixtures result in excessive leafiness.
Pruning Prune back well after flowering and tip prune new growth and young plants for more compact form.
Propagation Cuttings.
Possible Problems Apart from occasional leaf eaters is reasonably trouble free.

Goodenia ovata

Goodenia pinnatifida

Cut Leaf Goodenia,
Mother Ducks

FAMILY GOODENIACEAE

An appealing little plant for rock gardens or containers. The oblong to wedge-shaped leaves to 10 cm long are deeply toothed or with narrow lobes. The bright yellow flowers are carried on very fine branching stems, mostly from late winter to mid-summer.

Origin All States except Tas. and NT.
Size 10–30 cm high and as wide.
Soil Preference for gritty soils but adapts to most well-drained loams.
Climate Drier temperate to semi-tropical areas.
Aspect At least half sun.
Water Moderately; responds to water in late summer and autumn.
Feeding Slow release or complete plant food used sparingly when soil is moist during autumn.

Goodenia pinnatifida

Propagation Seed, cuttings or division.
Possible Problems Usually trouble free.

Grevillea

Some of the Hakea species are the only plants likely to be confused with Grevilleas. The main difference is in the seed capsules. Grevillea capsules (or follicles) are shell like, while those of Hakea are hard and woody, solid except for the centre seed cavity.

Grevillea candelabroides

Grevillea alpina

Grevillea alpina

(syn. *Grevillea dallachiana)*
FAMILY PROTEACEAE

A small shrub with many forms, most of them excellent for rock gardens or foreground situations with massed shrubbery. Leaves can be linear or oval and usually a grey-green. The flowers, which come in rather upright spidery clusters at ends of branchlets, may be cream and pink, yellow or frequently red and yellow and appear mainly during late winter, spring and summer.

Origin Vic., NSW, especially southern tablelands and ACT.
Size Most forms are about 60 cm high and are cushion shaped, however, some are prostrate and there are uprights to 2 m high. Width rarely more than 1–1.5 m.
Soil Well-drained and preferably crumbly loam.
Climate Cool to temperate; dislikes excessive humidity.
Aspect At least half sun.
Water Moderate to liberal.
Feeding Safer to use slow release or organic fertilisers.
Pruning Prune back as flowers finish.
Propagation Cuttings.
Possible Problems Psyllids or in wet humid areas fungus rot may cause dropping or blackening of flowers.

Grevillea banksii

FAMILY PROTEACEAE

An adaptable and long lived shrub with dark green leaves to 15 cm long, finely lobed almost to the midrib. Flower spikes are erect, cylindrical, to 18 cm long and are usually glossy red. The red form sold under the unofficial name *Grevillea banksii* 'Fosterii' has compact growth and flowers almost continuously.

Origin South-eastern Qld.
Size 2.5–3 m high and as wide. The white and some other wild forms are taller. There is also a prostrate form to 1 m high.
Soil Adapts to practically any but alkaline soils, even to wettish clay-loam. However if the latter is lightened by mixing in rotted leafmould and coarse sand, results are better.
Climate All but very frosty areas.
Aspect Accepts two-thirds shade but flowering and growth is better with at least half sun.
Water Moderate to liberal but fairly drought resistant when established.
Feeding Slow release or organic food may be applied sparingly in late spring to boost growth providing the plant is established and the soil is moist.
Pruning May be rejuvenated by moderate pruning; otherwise remove old flower spikes.
Propagation Cuttings; seed is productive but may be cross pollinated.
Possible Problems Except from damage by overfeeding, it is usually trouble free.

Grevillea bipinnatifida

FAMILY PROTEACEAE

Usually a sprawling shrub with attractively glossy, sharply and deeply cut, bipinnate leaves. Large light red tapering toothbrush-type flower spikes are displayed mainly during winter and spring then sporadically.

Origin WA, Darling Ranges and south.
Size Mostly about 0.5 m high and 1.5 m wide but there are more upright forms to 1 m high.
Soil From gravelly granite but adapts to most well-drained soils.
Climate Cooler temperate to semi-tropical.
Aspect At least half sun.
Water Moderately until established.
Feeding Sparingly if at all. with slow release or organic plant food.
Pruning Remove old flower heads. Shorten back branches to rejuvenate.
Propagation Cuttings and from seed.
Possible Problems Usually pest free but may succumb to fungus in humid areas.

Grevillea biternata

FAMILY PROTEACEAE

Initially a low carpeting plant with very finely divided light green glossy foliage. Flowering stems are erect to about 1 m or more, branching into rounded heads of fine whitish florets in rounded clusters with heavy honey aroma. These can be cut back in some forms to retain reasonably prostrate growth, others ascend to 1.5 m.

Origin WA, from Perth south-east to Stirling Ranges.
Size 1–1.5 m high to 2–3 m wide.
Soil Best in the gravelly soils of their natural habitat, at least a quick draining soil.
Climate Temperate; often short lived in high humidity.
Aspect At least half but preferably full sun.
Water Moderately until established.
Feeding If at all, sparingly with slow release or organic food.

Pruning Prune back ascending flower spikes when finished.
Propagation Cuttings.
Possible Problems Prone to die out in humid districts, especially in mass planting with restricted air circulation through the plants.

Grevillea buxifolia

Grey Spider Flower

FAMILY PROTEACEAE

A fairly sparse open shrub, the slender erect stems are clad with stiff oblong to oval leaves about 2 cm long and terminate in a rounded cluster of woolly grey florets during most of the year.

Origin NSW, especially in lightly timbered coastal sandstone regions.
Size 1–1.75 m high, rarely more than 1 m wide.
Soil At home in leaf mulched acid sandy areas but adapts to most well-drained loams.
Climate Cooler temperate to semi-tropical.
Aspect Light shade or broken sunlight to full sun.
Water Moderately.
Feeding A leafy mulch should be sufficient.
Pruning May be reduced to about one-

Grevillea biternata

Grevillea banksii

Grevillea bipinnatifida

Grevillea buxifolia

Grevillea x 'Canberra Gem'

third of its height, preferably without denuding it of foliage.
Propagation Seed or cuttings.
Possible Problems Occasionally develops mildew when too wet and sheltered.

Grevillea x 'Canberra Gem'

FAMILY PROTEACEAE

A shrub from the same parents as 'Pink Pearl' and almost identical to it. Apart from floral performance it is a good landscape subject for contrasting form and texture. Each erect branch is a well-defined dense pinnacle of deep green closely set foliage but still forming a compact rounded shrub. Clusters of bright pink and white flowers illuminate these erect branches during winter and spring.

Origin Hybrid of *Grevillea rosmarinifolia* and *G. juniperina*.
Size 2–2.3 m high, to 3 m wide.
Soil Very adaptable to most reasonably well drained garden soils.
Climate Cooler temperate to semi-tropical.
Aspect Best in full sun and well spaced to emphasise its form.
Water Moderate to liberal; accepts dryness when established.
Feeding Sparingly with slow release or organic food. Heavier feeding tends to yellow tips of foliage or result in more serious damage.
Pruning May be pruned back after flowering.
Propagation Cuttings.
Possible Problems Usually trouble free.

Grevillea candelabroides

FAMILY PROTEACEAE

An impressive grevillea with foliage resembling long erect pine needles divide into numerous needle-like segments. Held above this are long flower spikes in spectacular candelabra formation. These are initially glossy green in tight bud. They then change through lime to deep cream as they open, from early to late summer.

Origin WA, sandplain north of Darling Range.
Size 3–4 m high and nearly as wide.
Soil Very well-drained, composted sand or light gritty loam.
Climate Temperate to sub-tropical.
Aspect Open Sunny Situation.
Water Moderate only.
Feeding Slow release or organic fertiliser used sparingly in late autumn.
Pruning May be pruned back fairly heavily after flowering.
Propagation Seed.
Possible Problems Seems to be subject to root rot in humid east coast areas; possible a good candidate for grafting onto *Grevillea robusta* stock.

Grevillea x 'Crosbie Morrison'

Grevillea x 'Crosbie Morrison'

FAMILY PROTEACEAE

An adaptable and attractive small dense foliage shrub, stems covered with closely set needle like leaves, greenish above, grey on underside and along branchlets, giving an overall greyish appearance. During late winter and spring, spidery clusters of crimson flowers are carried- along the branches, often heavily concentrated towards the tips giving a showy bunch like appearance.

Origin Hybrid from *Grevillea lanigera* and *G. lavandulaceae.*
Size 1–2 m high and similar width.
Soil Adapts to a wide range of well-drained soils.
Climate Cooler temperate to semi-tropical.
Aspect Best with at least two-thirds sun.
Water Moderate to liberal.
Feeding Sparingly with slow release or organic foods.
Pruning For more compact growth prune back as flowers finish.
Propagation Cuttings.
Possible Problems Borer attack possible in some areas.

Grevillea dielsiana

FAMILY PROTEACEAE

An unusual small shrubby grevillea with cylindrical foliage no thicker than a fine knitting needle, divided at least three times into shortening and eventually sharp segments. The glossy, slightly pendulous spring and early summer flowers are in loose sprays, one form bright scarlet and the other bright yellow.

Origin WA, for about 100 km north of Perth.
Size Average growth about 0.3 m per year; to 1–1.5 m high, to 1.5 m wide.
Soil Well-drained and preferably crumbly or gritty soil.
Climate Temperate to semi-tropical.
Aspect Best in open sunny situation.
Water Moderately, at least until established.
Feeding Needs little more than a leafy mulch; slow release used sparingly may speed growth.
Pruning Prune lightly to remove old flower sprays and for more compact growth.
Propagation Cuttings or seed.
Possible Problems Overfeeding can injure the plant.

Grevillea dryandri

FAMILY PROTEACEAE

A low spreading plant with near horizontal growth and very finely

Grevillea dielsiana

Grevillea dryandri

divided leaves to about 20 cm long. The extremely long and slender toothbrush type flower sprays are also nearly parallel to the ground, mostly with lavender to pink and red florets extending for about 40 cm. Different forms vary from white to bright red. Flowers from mid-summer to autumn.

Origin Northern Qld, NT and into WA.
Size 30–40 cm high, spreading about 2 m.
Soil Seems to favour slightly clayey loam, subject to some wet season inundation, but also accepts good drainage.
Climate Tropical to warmer temperate.
Aspect Usually found in broken sunlight.
Water Liberal water during summer.
Feeding Slow release or organic food in summer when soil is moist.
Pruning Prune lightly after flowering.
Propagation Cuttings or seed if away from cross-pollination.
Possible Problems Usually trouble free in tropical areas with dry winters but difficult to establish in many areas unless grafted onto *Grevillea robusta* stock.

Grevillea x gaudichaudii

Grevillea eriostachya

Honey Grevillea

FAMILY PROTEACEAE

A most spectacular spreading shrub for the drier inland areas with long branches ascending at about 45°. These are veiled in leaves to 20 cm long, which are divided into three or five wiry segments and terminate in a branching inflorescence of up to seven tapering spikes 12–18 cm in length. They are at first woolly grey, then bright green, before opening into glossy golden florets.

Grevillea eriostachya

Origin From the central coast of WA into central Australia and western Qld. The one illustrated was photographed close to Ayers Rock.
Size 3–4 m.high and as wide.
Soil Well-drained, preferably gritty loam.
Climate Drier temperate and semi-tropical.
Aspect Fairly open position; accepts exposure and wind, although for culture nearer to Perth, wind protection is recommended.
Water Moderate only; native habitats receive less than 250 mm (10 in) of rain per year.
Feeding Organic mulch or light application of slow release fertiliser.
Pruning Branches may be pruned back by at least half after flowering.
Propagation Seed or cuttings.
Possible Problems Rarely affected, at least not in well-drained low rainfall areas.

Grevillea x *gaudichaudii*

FAMILY PROTEACEAE

A popular and attractive ground cover for non-humid and relatively cool districts. It has similar appearance to *Grevillea* X 'Royal Mantle' but the 15–18 cm long leaves have five to seven definitely sharp pointed lobes, with some resemblance to a slender oak leaf. Dark red toothbrush-type flower clusters adorn the foliage mat during winter and spring. A popular bank cover in Canberra and parts of Vic.

Origin Natural hybrid from the Blue Mountains (*Grevillea laurifolia* X *acanthifolia*).
Size 10–12 cm high, with a spread 2–3 m.
Soil Well- drained sandy loam or if heavier clay, well broken by mixing in rotted leafmould or compost.
Climate Cooler and drier temperate areas rather than humid coast.
Aspect Ideal in dappled sunlight but will accept more sun.

Water Moderate.
Feeding Sparingly with organic or slow release fertiliser.
Pruning May be pruned back lightly to thicken.
Propagation From cuttings.
Possible Problems Overfeeding causes yellowing foliage. Also thrips may cause silvery mottling, especially where plants have overhead protection or during very dry conditions.

Grevillea hookeriana

Grevillea hookeriana

FAMILY PROTEACEAE

The grevillea widely sold under this name differs greatly from the original south-western Australian species both in foliage and flower. However the usurper is an appealing and a useful screen plant. Foliage is bright green, to about 15 cm long and divided into herringbone formation. Toothbrush flower spikes, 5–8 cm long, gain their brilliant red mostly from the long glossy pendulous styles that almost veil the brush. These showy flowers are carried through most of the year.

Origin Unknown; believed to be a hybrid.
Size 3–4 m high and as wide.
Soil Adapts to a wide range including moist clay.
Climate Cooler temperate to semi-tropical; good frost resistance.
Aspect At least half sun; some shelter from cold winds in cool areas.
Water Moderate, at least until established.
Feeding If at all, slow release or organic food, only when soil is moist.
Pruning May be pruned back heavily preferably in late spring or autumn when growth is relatively inactive.
Propagation Cuttings.
Possible Problems Relatively free except perhaps damage from overfeeding or, in some areas, borer.

Grevillea x *Honey Gem*

Grevillea x 'Honey Gem'

FAMILY PROTEACEAE

A tall growing grevillea resembling its *Grevillea pteridifolia* parent. It has similar large fine fishbone-type leaves to about 30 cm long, perhaps grey-green rather than silver-grey, and a little more pendulous, giving a denser effect. Also the large flower brushes are the similar one-sided toothbrush-type 15–17 cm long, honey coloured with a touch of *G. banksii* reddish brown at the base rather than overall golden-orange. It is long flowering, mainly in autumn, winter and spring.

Origin Hybrid from two north Australian species *G. banksii* X *G. pteridifolia*.
Size 3–4 m high, 2.5–3 m wide.
Soil Adapts to a wide range of reasonably well-drained soils.
Climate Temperate to tropical but is unlikely to have much resistance to heavy frost.
Aspect Fairly open sunny position.
Water Moderate. Good drought resistance when established.
Fertiliser Light application of slow release or organic fertiliser, but only while soil is moist.
Pruning: May be heavily pruned if necessary.
Propagation Seed.
Possible Problems As far as known at time of writing this grevillea is fairly trouble free, except perhaps when soils remain wet for a long period, especially during cold conditions.

Grevillea x 'Ivanhoe'

Grevillea x 'Ivanhoe'

FAMILY PROTEACEAE

An adaptable and fast growing tall shrub excellent for quick screening or an attractive background. Slender herringbone-type leaves are about 15 cm long with only short lobes bronze red when new, then mid to deep green. Light purple brushes 5-6 cm long are carried both at the tips and along the branches mainly from mid-winter to late spring.

Origin Hybrid of uncertain parentage.
Size 3-5 m high, 2-3 m wide.
Soil Grows well in a variety of soils including frequently wet sticky clay soil.
Climate Cooler temperate to tropical.
Aspect As little as one-third to full sun.
Water Liberal to moderate with reasonable drought resistance.
Feeding Very sparingly if at all with organic or slow release but leafy mulch or compost is usually enough.
Pruning May be pruned heavily to rejuvenate or clipped as a hedge.
Propagation Cuttings.
Possible Problems Because of rapid growth can become top heavy and blow over. Pruning back young plants to encourage more base branching should help, even though it may appear to retard initially. Also check for borer especially in spring.

Grevillea johnsonii

Grevillea johnsonii

FAMILY PROTEACEAE

An erect rarely dense but compact shrub attractive mainly for its dark green pendulous foliage. Leaves are to 20 cm long and although divided pinnately the segments are long and thin, giving a slightly pendulous pine-like effect. The individual florets are shrimp-like, red and white on short spikes along newer growth during spring. There seems some confusion between this species and *Grevillea longistyla* from Queensland. However this illustration is of a specimen confirmed as *G. johnsonii* in Canberra Botanic Gardens.

Origin Tablelands and slopes of NSW.
Size 4-5 m high, about 3 m wide.
Soil Occurs particularly in rocky areas but adapts to most well-drained soils.
Climate Cooler temperate to temperate; unreliable on humid coast unless grafted on *G. robusta*.
Aspect Dappled to full sunlight.
Water Moderate to liberal but reasonable drought resistance once established.
Feeding Sparingly; see *G. hookeriana*.
Pruning May be rejuvenated by heavy pruning after flowering.
Propagation Cuttings.
Possible Problems As *G. hookeriana*.

Grevillea juniperina

FAMILY PROTEACEAE

Like some other grevillea species, this one comes in several forms. All have attractive, but prickly, stiff, dark green slender or almost needle-like linear leaves to about 2 cm long and semi-erect spidery clusters of flowers from early winter to spring. Upright forms have either yellow or red flowers; prostrate forms are yellow, red or pink. They all display their flowers above the foliage. 'Mongolo' (illustrated) is a hybrid between the upright red and prostrate yellow forms and has apricot flowers and greater vigour. There is also a variegated form called 'Lunar Light'.

Origin NSW tablelands and coast.
Size Upright forms to 2 m and nearly as wide, prostrate to about 30 cm spreading to 2 m following contours.
Soil Any average garden soil not heavily limed or fertilised recently.
Climate Cooler temperate to semi-tropical; good frost resistance.
Aspect Best with at least two-thirds sun.
Water Liberal to moderate; reasonable drought resistance.
Feeding Sparingly if at all; see *Grevillea hookeriana*.
Pruning Upright forms especially may be pruned back heavily in spring.
Propagation Cuttings.
Possible Problems Good resistance to root rot except in overwet soils; mites may mottle foliage in dry climates especially when grown with overhead cover.

Grevillea lanigera

(prostrate form)
FAMILY PROTEACEAE

Worth growing for its ornate foliage alone. The stout but prostrate wiry stems are almost hidden by a very closely set spiralling arrangement of deep to silvery green linear leaves, to about 2 cm long and neatly rounded at the tip. Minute hairs give a satin-like appearance. Clusters of shrimp red and cream flowers add colour during winter and spring.

Origin Vic. and NSW.
Size Comparatively slow growing, to about 40 cm high and spreading 1–2 cm.
Soil Well-drained slightly acid sandy or medium loam.
Climate Cooler temperate to semi-tropical.
Aspect At least half sun.
Water Moderate, at least until established.
Feeding Light dressings of slow release or organic fertilisers in spring will speed growth once plants are established.
Pruning May be pruned after flowering.
Propagation Cuttings.
Possible Problems Under foliage may tend to brown in wet or humid areas. Growing on a pebbly scree or over larger stones seems to minimise this.

Grevillea juniperina (prostrate yellow form)

Grevillea lanigera (prostrate form)

Grevillea lavandulaceae

Grevillea lavandulaceae (prostrate)

Grevillea lavandulaceae

FAMILY PROTEACEAE

There are many forms, shrubby with unusually straight and almost erect stems providing interesting contrast amongst mixed shrubbery, others with scrambling semi-prostrate growth. All have slender foliage 1–2 cm long, sharply pointed at the tip rather than rounded and profuse clusters of spidery flowers all along the branches, mainly from mid-winter to late spring; colour mostly red but there are also pink and white forms.

Origin SA, Vic. and NSW.
Size Upright form to 2 m high and nearly as wide, semi-prostrate about 0.5 m high and 1.5–2 m wide.
Soil Adapt to a wide range, including heavy clay and limey soils providing they are not sedgy.
Climate Cooler temperate to semi-tropical.
Aspect Best in an open sunny position.
Water Moderate to liberal but good drought resistance.
Feeding As *Grevillea lanigera*.
Pruning Upright types especially may be pruned back hard after flowering; shorten back semi-prostrate forms for bushier growth.
Propagation Cuttings.
Possible Problems Good resistance to root-rotting fungi but check in spring for borer.

Grevillea longifolia

FAMILY PROTEACEAE

An interesting large spreading shrub with long slender, but rigid, deep green saw-toothed leaves to 25 cm long, held nearly horizontal rather than pendulous, and on a similar plane, bright rosy red toothbrush-type flower spikes, 6–10 cm long, appear profusely during winter and spring but also sporadically throughout the year.

Origin NSW, mostly in lightly sheltered mountain and coastal gullies.
Size 2–2.5 m high and to 4 m wide.
Soil Adapts to most, including badly drained garden soils, but will not tolerate heavily limed soils.
Climate Cooler temperate to semi-tropical with some shelter from cold winds (growing well under broken sunlight in Canberra).
Aspect Seems best with lightly dappled shade to half sun.
Water Moderate to liberal; good resistance to root rot and, once established, to dry conditions.
Feeding No more than sparingly with slow release or organic food.
Pruning May be pruned back hard to rejuvenate.
Propagation Cuttings.
Possible Problems Fairly long lived but check occasionally for signs of borer damage.

Grevillea x 'Misty Pink'

FAMILY PROTEACEAE

An appealing shrub for any type of garden in all but very frosty areas. The broad leaves with their long slender lobes are similar to those of parent *Grevillea banksii* but are slightly silvery rather than dark green. Flowers are also in cylindrical spikes but these are up to 15 cm long and soft pink with cream styles. These are offering almost throughout the year, more profusely during spring and again in autumn.

Origin Hybrid between *G. banksii* and *G. sessilis*.
Size 2–3 m high, to about 2 m wide.
Soil Poor sandy soil to puggy clay.
Climate All but cold frosty areas.
Aspect At least half sun.
Water Moderate to liberal. One near Sydney has been growing and flowering

Grevillea x 'Misty Pink'

Grevillea longifolia

well for over five years in poor sand within 50 cm of the base of a sixty year old jacaranda, depending only on rainfall.
Feeding A little compost or leafy mulch would be appreciated.
Pruning Remove seed heads or prune well back into old wood to rejuvenate.
Propagation Cuttings.
Possible Problems Surprisingly trouble free. Reputed danger of wind damage can be minimised by pruning hard to encourage a low branching and self-bracing-type growth, at the same time avoiding heavy feeding.

Grevillea muelleri

FAMILY PROTEACEAE

An interesting and different little grevillea suitable for rock gardens or even possibly for large containers. It makes a spreading clump of erect reddish stems with pine-like bright green foliage also erect. A profusion of creamy yellow flowers in globular clusters give a pleasing display during winter and early spring.

Origin WA, apparently in the south-west.
Size About 50 cm high, to 1 m across.
Soil At time of writing not widely grown but is doing well in well-drained situations in Melbourne and in a mildly acid sandstone area near Sydney.
Climate At least suitable for cooler temperate to temperate areas.
Aspect In both above areas dappled sunlight seems ideal.
Water Moderate, or to liberal if drainage is good.
Feeding Very sparingly, perhaps a sparing dressing of slow release.
Pruning Should respond to moderate pruning after flowering.
Propagation Cuttings.
Possible Problems Seems trouble free under conditions suggested.

Grevillea 'Pink Pearl'

See *Grevillea* 'Canberra Gem'

Grevillea 'Poorinda Illumina'

FAMILY PROTEACEAE

A small shrub with fairly dense but very slender short (1-2 cm long) grey foliage, valuable for contrast amongst other plants as well as for the bright pink and cream flowers clustered over it during winter and spring.

Origin Hybrid between *Grevillea lavandulaceae* and *G. lanigera*.
Size 1-1.7 m high, to about 1.5 m wide.
Soil Adaptable to most reasonably drained soils but seems best in gravelly loam or at least a stony surface scree.
Climate Cool and temperate.
Aspect Lightly broken to full sunlight.
Water Moderate; withstands dryness when established.
Feeding Slow release or organic food used sparingly.
Pruning May be pruned back to about half after flowering.
Propagation Cuttings.
Possible Problems Fairly trouble free but occasional borer attack possible.

Grevillea 'Poorinda Royal Mantle'

FAMILY PROTEACEAE

One of the most widely used groundcover plants during the seventies and eighties. It makes a low and dense contour-following mat of slender irregularly shaped leaves to about 14 cm long, some entire or plain, others unevenly lobed. New growth is bronze-purple. Raspberry red to pale purple toothbrush-type flowers rest slightly above the foliage, particularly during

Grevillea muelleri

Grevillea 'Poorinda Illumina'

Grevillea 'Poorinda Royal Mantle'

summer. This grevillea is also grafted at various heights onto *Grevillea robusta* stock to feature as a weeping standard.

Origin Hybrid between *G. laurifolia* and *G. willisii*.
Size Stems mat to about 15 cm high, spreading 2-3 m, sometimes eventually to 5 m.
Soil Very adaptable, both to quick drying sandy or heavy moist clay soil.
Climate Cool, temperate to tropical.
Aspect Needs at least two-thirds sun to flower well but accepts light shade.
Water Moderate to liberal; accepts dry periods fairly well except when fertiliser has been used nearby.
Feeding Very sparingly if at all with slow release or organic food.
Pruning Not necessary except to restrain spread.
Propagation Cuttings.
Possible Problems Yellowing foliage or sometimes greater damage from heavy fertilising near root area. Weeds can be a problem when establishing as a groundcover unless soil is blanketed with mulch.

Grevillea robusta

Silky Oak

FAMILY PROTEACEAE

A giant amongst the grevilleas, more suitable for street or park planting than for most home gardens. Its attractively divided fern-like foliage makes it a prized pot plant in the northern hemisphere where it is too cold for its normal growth,and during late spring or early summer the masses of golden orange flower brushes suggest a giant torch in the landscape. It is a paradise for birds, a valuable timber tree and its seedlings are a useful grafting stock for difficult to grow grevilleas otherwise subject to root rot.

Origin Rainforest of Qld and northern NSW.
Size 20–30 m high, pyramid shape, about 10 m wide.
Soil Adapts to most well-drained soils.
Climate Temperate to tropical; reasonable frost tolerance when established.
Aspect May start life in shade and shelter; needs full sun to flower well.
Water Liberal, especially in summer but drought resistance good.
Feeding Cow manure lightly mulched or slow release food once established will speed growth.
Pruning Better left to grow naturally.
Propagation Seed.
Possible Problems Usually fairly trouble free for the first fifty to sixty years of growth; then may succumb to fungus disease and borer.

Grevillea 'Robyn Gordon'

FAMILY PROTEACEAE

The low growing, compact habit, attractive foliage and almost perpetual

Grevillea robusta

Grevillea 'Robyn Gordon'

flowering has given this Australian raised grevillea great popularity in gardens throughout the world. The 12–15 cm glossy bright green leaves are ornately divided into numerous trident or double trident lobes and the gracefully tapering one-sided flower spikes are at least as long, light red and glossy textured. These are most prolific during autumn, winter and spring but at least some are also displayed during most of the summer. Good tub or rock garden plant.

Origin Hybrid between *Grevillea banksii* and *G. bipinnatifida.*
Size About 1 m high, 1.5–2 m wide.
Soil Adapts to most well-drained soils.
Climate Performance more outstanding in warmer temperate to semi tropical areas but has good frost resistance and is worthwhile in cooler temperate districts.
Aspect Preferably full sun but accepts part shade in warmer districts.
Water Moderate to liberal; withstands dryness when established.
Feeding Sparingly with slow release or organic food; only when soil is moist.
Pruning Best when pruned back fairly heavily, preferably when spring flower peak subsides.
Propagation Cuttings.
Possible Problems Some spotting and leaf shrivelling due to fungus attack sometimes occurs during cool conditions. Liberal fertilising causes similar yellowing. Also check occasionally for borer.

Grevillea rosmarinifolia

Grevillea rosmarinifolia

FAMILY PROTEACEAE

At one time one of the few grevilleas grown in home gardens and the best known. There are now more spectacular flowers but it is still an attractive shrub or screen plant with dense bright to deep green needle foliage 3–4 cm long. Flowers are rose pink to red with a creamy white blotch, occurring mainly from late autumn to mid-spring. There are various forms, especially dwarf types but most only perform well in certain areas.

Origin Vic. and NSW.
Size Original tall form 1.5–2.5 m high, 2–3 m wide.
Soil Adapts to a wide range of well-drained soils.
Climate Cooler temperate to semi-tropical; many dwarf forms more suited to drier temperate areas.
Aspect At least half sun.
Water Moderate; resistant to dry periods when established.
Feeding Sparingly with slow release or complete organic foods.
Pruning May be rejuvenated by heavy pruning; it is a good hedge if pruned regularly.
Propagation Cuttings.
Possible Problems Borer, especially in mild coastal areas. Psyllids may cause bud drop.

Grevillea x 'Sandra Gordon'

FAMILY PROTEACEAE

A fairly sparse small tree or, with heavy pruning, a dense shrub. The leaves, from 20–25 cm long are finely divided almost to the midrib into nine to twenty slender alternate linear lobes 6–12 cm long, fairly deep green above but silvery grey at the back, the latter emphasising and increasing the wispy appearance when allowed to grow tall naturally. Spectacular from autumn to spring with 15 cm long one-sided bright yellow nectar-laden flower spikes held high.

Origin Hybrid between *Grevillea pteridifolia* and *G. sessilis.*

Grevillea x 'Sandra Gordon'

selections are available from most native plant nurseries. A most adaptable species.

Origin NSW, particularly in coastal Hawkesbury sandstone and Narrabeen shale areas.
Size To 2 m high, 1-1.5 m wide; more compact forms available.
Soil Sand or loam preferably with a leafy mulch and reasonable drainage.
Climate Temperate to semi-tropical; good frost tolerance.
Aspect Dappled to full sunlight.
Water Moderate to liberal; accepts dry periods when established.
Feeding A light application of slow release or organic plant food is acceptable in spring.
Pruning Needed to keep compact form, mainly after main spring flower flush then tip prune.
Propagation Cuttings.
Possible Problems Occasionally visited by a tip-rolling caterpillar which performs a clumsy type of tip pruning.

Size 5-7 m high or with pruning kept to about 3 m, natural width 3-4 m.
Soil Accepts heavy or even moist clay loam but a gritty well-drained loam seems preferable.
Climate Temperate to tropical; reasonable frost tolerance.
Aspect At least half but preferably full sun.
Water Moderate; good drought tolerance when established.
Feeding Sparingly with slow release or organic food.
Pruning For compact shrubby growth prune back heavily in spring when flowers finish.
Propagation Seed.
Possible Problems Deforming and small root-like clusters on stem usually due to mite or psyllid injury, particularly when sheltered from weather.

Grevillea sericea

Pink Spider Flower

FAMILY PROTEACEAE

Usually a light open, erect shrub, leaves slender elliptical to linear, 3-5 cm long, light green in colour. Flowers are pink, in spidery clusters 3-4 cm across during most of the year. Form, flower colour and size vary in different localities but good compact and deeply coloured

Grevillea sericea

Grevillea x 'Shirley Howie'

Grevillea thelmanniana (upright form)

Grevillea x 'Shirley Howie'

FAMILY PROTEACEAE

Resembles a better form of one parent, *Grevillea sericea*, but flowers are larger, deeper pink. Also the slender elliptical to lance-shaped lightish green foliage is glossier and much denser. A worthwhile attractively shaped shrub, flowering mainly during winter and spring.

Origin Hybrid between *G. capitella* and *G. sericea*.
Size 1.5–1.7 m high and almost as wide.
Soil Adapts to most well-drained soils. Prefers a fairly crumbly slightly acid loam.
Climate Temperate to tropical. Grows well in Brisbane gardens.
Aspect At least half sun.
Water Moderate to liberal.
Feeding Sparingly only, with slow release or organic.
Pruning Prune moderately to heavily as flowers finish in spring.
Propagation Cuttings.
Possible Problems Fairly trouble free.

Grevillea tetragonaloba

Grevillea tetragonaloba

FAMILY PROTEACEAE

The cumbersome species name merely suggest 'four-angled lobes'. A low spreading shrub with the rather dense foliage divided into three to five lobes of irregular length which with its slightly grey-green colour gives a lacy appearance. The florets, tightly packed in toothbrush form to 10 cm long are pinkish white with long red styles, suggesting raspberry coconut bars or similar confection. These appear throughout the year, more profusely in autumn and spring. Also known as *Grevillea tenuiloba*, a name which seems more correctly applicable to a pinnate leafed form of *G. tetragonaloba*.

Origin WA.
Size 1 m high, to about 2 m across (under suitable conditions)
Soil Well-drained, preferably gritty and no more than slightly acid.
Climate Temperate; more suited to relatively dry rather than east coast summers. Moderate frost tolerance.
Aspect Lightly broken or half to full sunlight.
Water Moderately; subject to root rot in moist humid areas unless grafted on

Grevillea triloba

G. robusta as the one illustrated.
Feeding Sparingly, with slow release or organic food.
Pruning May be shortened back to rejuvenate or for bushier growth.
Propagation Cuttings, apparently with some difficulty.
Possible Problems Root rot in humid areas unless grafted on to *G. robusta* stock.

Grevillea thelmanniana

(upright grey foliage form)
Spider Net Grevillea
FAMILY PROTEACEAE

A lovely grevillea with nearly erect stems densely clad in very finely divided but erect, almost wire-like, silver-grey to grey-green foliage on short silvery branchlets. The latter terminate in slightly nodding bright red toothbrush-type flower spikes 6–8 cm long during winter and spring. There is also a more adaptable upright green foliaged form.

Origin WA, south coast (Warren and Eyre regions).
Size 1–1.5 m high and as wide.
Soil Needs particularly good drainage, preferably a gritty limey soil, otherwise a pebbly scree at the base should help.
Climate Temperate; resents high humidity and heavy frosts.
Aspect Sun or very lightly broken sunlight and good air circulation.
Water Moderate only; once established allow soil surface to dry for a few days between waterings.
Feeding Sparingly, with organic or slow release in spring.
Pruning After flowering stems may be reduced to about half.
Propagation Cuttings.
Possible Problems Usually short lived in wet summer areas or acid soils; adding limestone chip may help.

Grevillea thelmanniana var. *Pressii* (prostrate)

Grevillea thelmanniana var. *Pressii*

FAMILY PROTEACEAE

A most attractive prostrate shrub with dense and very finely divided foliage creating a soft fluffy appearance. These fine fishbone-type leaves are deep green with a hint of blue-grey and form a delightful background during winter and spring for the bright red short toothbrush-type flower clusters. There is also a grey-foliaged prostrate form, but it is less adaptable than this one.

Origin Southern WA.
Size 30–35 cm high, spreading 1–1.5 m.
Soil All but very acid or heavy clay loams providing drainage is good.
Climate Temperate to semi-tropical but not severe frosts.
Aspect At least two-thirds sun.
Water Moderately, at least until established.
Feeding Slow release or organic food acceptable in spring when established.
Pruning If necessary shorten back to encourage bushiness.
Propagation Cuttings.
Possible Problems Comparatively trouble free.

Grevillea triloba

FAMILY PROTEACEAE

A medium sized shrub with fairly dense prickly foliage as each rigid grey-green leaf is divided into three lance-like segments. The very fine creamy white honey-scented flowers come in feathery globular clusters, almost covering the bush during winter and early spring.

Origin WA.
Size Height to 2 m and nearly as wide but can be confined by pruning.
Soil Adapts to both heavy and light soils with good drainage.
Climate Cooler temperate to semi-tropical.
Aspect Preferably full sun.
Water Moderate to liberal; good drought resistance.
Feeding Sparingly with organic or slow release.
Pruning May be kept compact by pruning back after flowering then tip pruning new growth.
Propagation Cuttings.
Possible Problems Usually trouble free in well drained soils without excessive fertilisers.

Gossypium australe

FAMILY MALVACEAE

A low bushy or procumbent plant with coarse textured narrowly or broadly oval (occasionally lobed) foliage; young shoots hairy, flowers 5–7 cm across, mauve to pink with a prominent dark red blotch at the base of each petal; flowers mainly during winter and spring but sporadic.

Origin WA, NT and western Qld.
Size 0.2–1 m, spreading 1–1.5 m.
Soil Preference for gravelly well-drained soil.
Climate Mainly temperate to tropical semi-arid regions; frost sensitive but new base shoots usually follow frost burn.
Aspect Full sun to lightly broken sunlight.
Water Moderate until established; good drought resistance.
Feeding Light application of complete fertiliser or fowl manure in spring if soil is moist.
Pruning May be pruned back as flowers finish to encourage bushiness.
Propagation Seed.
Possible Problems Weevils or small beetles may leave small holes in foliage and also attack flower buds.

Gossypium australe

Gossypium sturtianum

(syn. *Gossypium sturtii*)

Sturt's Desert Rose

FAMILY MALVACEAE

A small soft wooded erect shrub with reddish brown stems, rounded leaves 5–7 cm long, smooth deep green above but as they are usually rolled they display mainly their pale grey-green underside. The flowers, like small hibiscus, are either blush pink, pink or soft lilac, in all cases with a dark red centre blotch highlighting the white stamens. These occur throughout the year.

Gossypium sturtianum

Origin The arid areas of all Australian mainland States, especially Central Australia.
Size 1–2 m high, to 1 m wide.
Soil Very well-drained, preferably gritty loam.
Climate Drier temperate to tropical regions.
Aspect Full sunlight and open position with good air circulation.
Water Moderately only until well established then let soil dry for at least a few days before watering again.
Feeding Light dressing of complete slow release or fowl manure but then do not allow complete dryness of soil for a week or two.
Pruning May be rejuvenated by pruning back fairly heavily in spring.
Propagation Seed.
Possible Problems Flowers and foliage may be attacked by weevils or beetles, also by fungus in humid areas.

Hakea

Some of the Grevillea species are the only plants likely to be confused with Hakeas. The main difference is in the seed capsules. Grevillea capsules (or follicles) are shell like, while those of Hakea are hard and woody, solid except for the centre seed cavity.

Hakea bakerana

FAMILY PROTEACEAE

A medium to small woody but attractive compact shrub with slender midgreen needle-like leaves to 7 cm long and dense clusters of pink flowers with long thin styles are carried on the older branches during winter and spring.

Origin NSW central coast.
Size 2.5–3 m high, about 2 m wide.
Soil Well-drained composted sand or light to medium loam.
Climate Temperate to semi-tropical.
Aspect Lightly broken to full sunlight.
Water Moderate to liberal.
Feeding Slow release or organic food in spring.
Pruning May be pruned back after flowering if necessary.
Propagation Seed or cuttings.
Possible Problems Seems trouble free, at least in coastal areas.

Hakea bucculenta

FAMILY PROTEACEAE

A tall and erect shrub making a gentle screen of fine linear to flattened pine-like leaves about 15 cm long and from late

Hakea bakerana

winter to mid spring decked with showy erect tapering and densely packed orange-red flower spikes to 13 cm long.

Origin WA, sandplain north of Darling Range and extending east towards north of Coolgardie.
Size 4–5 m high, 2–2.5 m wide.
Soil Slightly limey well drained light loamy soil or gritty well mulched composted sand.
Climate Temperate to semi-tropical with fairly dry summer. Frost resistant.
Aspect In native habitat grows in open sun amongst low scrub but also growing happily with dappled shade at Canberra.
Water Moderately only until established; resents frequent watering during summer.
Feeding If any, only light application of slow release or organic fertiliser as new growth commences.
Pruning May be pruned moderately after flowering.
Propagation Seed.
Possible Problems Root rot in wet humid summer conditions.

Hakea bucculenta

Hakea chordophylla

Hakea chordophylla

FAMILY PROTEACEAE

A small tree with brown, deeply fissured corky bark and smooth, grey, cylindrical almost wire-like (terete) leaves from 15–40 cm long and erect brushes, 10–14 cm long of glossy looking cream flowers, mainly during winter and spring.

Origin Central Australia, mainly lower NT and spreading into WA and Qld, mainly on sandplains. Occasionally near river banks with seasonal inundation.
Size 3–7 m high and 3–5 m wide.
Soil Slightly limey to no more than mildly acid deep, gritty sand with good drainage.
Climate Temperate to drier tropical; seems frost tolerant and should suit drier inland districts.
Aspect Full sun.
Water Moderate only until established. Good drought resistance.
Feeding Sparingly with organic fertilizer, if at all.
Pruning: Rarely necessary except perhaps for shape.
Possible Problems: Seems trouble free in its natural environment.

Hakea laurina

Hakea eyreana

Hakea multilineata

Hakea eyreana

(syn. *Grevillea eyreana, Hakea divaricata, H. intermedia*)

FAMILY PROTEACEAE

An unusual small tree with a brown, deeply furrowed trunk and lower branches supporting what from a distance looks like a large undulating dark green net. The leaves are needle-like (terete), up to 16 cm long and divided three to fifteen times, the final divisions sharply pointed. The racemes of spring flowers are from 8–14 cm long, glossy green in bud, deepening to lime then opening yellowish cream, sometimes terminal but also bursting from main limbs or even from the bark-covered trunk.

Origin Central Australia and about one-third of way up NT and across the border of WA and Qld, mainly on flats with occasional inundation. There is also a slightly hairy rather than glossy flowered form extending to just south of Lake Eyre.
Size 5–7m high, 3–4 m wide.
Soil At home in well-drained red sand, or else a gritty gravelly loam.
Climate More arid temperate to semi-tropical.
Aspect Open sunny situation.
Water Moderate but accepts inundation for up to three or four weeks and up to six months dryness.
Feeding Sparingly only with slow release or organic complete.
Pruning Should not be necessary except perhaps to rejuvenate.
Propagation Seed.
Possible Problems Rarely affected.

Hakea laurina

Pincushion Hakea

FAMILY PROTEACEAE

A lovely hakea, developing to a large erect shrub or small tree with bluish green slender elliptical leaves to about 15 cm long; young growth bronze. During autumn and winter it carries eye-catching globular heads of reddish flowers lightly veiled by pin-like white styles. Long lasting as cut flowers.

Origin WA, south coast.
Size 4–6 m high, 2.5–3 m wide but occasionally wider and shorter.
Soil Well-drained light or stony soil

Hakea oleifolia

preferable but has also proved successful in heavy clay lightened by leafmould.
Climate Cooler temperate to semi-tropical; good frost tolerance and being from a partly summer rainfall area accepts east Australian coastal conditions.
Aspect At least half sun.
Water Moderate to liberal but reasonable drought resistance when established.
Feeding Sparingly with organic or slow release fertiliser in spring.
Pruning May be moderately pruned before new growth starts.
Propagation Seed.
Possible Problems Scale occasionally occurs, causing sooty mould on foliage or stems. Wind damage when top heavy — avoid this by starting with young unstaked plant.

Hakea multilineata

Grass-leafed Hakea

FAMILY PROTEACEAE

Upright shrub with erect slender oblong to linear grey-green leaves to about 20 cm long and displaying numerous parallel veins (as suggested by species name). Scarlet flowers are clustered towards the tips of the erect branchlets in spikes to 15 cm long during late winter and spring. There is also a white form. Popular garden plant around Perth and Adelaide.

Origin WA, mostly south of Stirling Ranges but also some areas north.
Size 4–6 m high, to 3 m wide.
Soil Well-drained, slightly limey, light and preferably gritty loam.
Climate Temperate to semi-tropical; good frost tolerance.
Aspect Open sunny situation. In windy areas start with young plant and cut back to encourage self-bracing low branching.
Water Moderate; good drought resistance when established.
Feeding Well-composted soil should be sufficient, otherwise sparing use of slow release or organic fertiliser in spring.
Pruning May be pruned back in spring after flowering.
Propagation Seed.
Possible Problems May be short lived on humid east coast.

Hakea oleifolia

Olive Leaf Hakea

FAMILY PROTEACEAE

Actually the leaves are a little broader and rounder at the tip than olive leaves and a pleasant blue-grey-green on both sides. Growth is dense and bushy; good background or windbreak. Globular clusters of white flowers are prominent above the foliage in mid-spring.

Origin Lower south-western Australia.
Size 5–6 m high, 3–4 m wide.
Soil Preference for sandy soil but accepts well-drained loam.
Climate Cooler temperate to temperate; frost resistant.
Aspect Best in full sunlight.
Water Moderate to liberal but good drought resistance.
Feeding Sparingly with slow release or organic fertiliser when established.
Pruning May be pruned back as flowers finish or in earlier growth to encourage bushiness.
Propagation Seed.
Possible Problems May not adapt well to some hot humid east coast areas.

Hakea plurinervia

FAMILY PROTEACEAE

An erect and clean cut shrub, smooth brown branchlets with soft green lance-shaped leaves to 15 cm long, broadening below the tips, and in spring, dense globular clusters of white flowers in leaf axils.

Origin Qld.
Size 2–3 m high, 1–2 m wide.
Soil Adapts to most well-drained soils.
Climate Cooler temperate to tropical; good frost tolerance.
Aspect Full to lightly broken sunlight.
Water Moderate to liberal but when established has reasonable drought resistance.
Fertiliser Organic or slow release plant food in spring.
Pruning May be pruned back after flowering.
Propagation Seed.
Possible Problems In some areas scale — presence usually indicated by black sooty mould.

Hakea prostrata

(syn. *Hakea glabella*)

FAMILY PROTEACEAE

A hakea which often belies its name because it is usually a tall bushy shrub. Possibly the prostrate form was the first named. The clusters of creamy white flowers with purple stigmas lie snugly in the axils of the light green slightly cupped oval leaves with rather sharp toothed margins.

Origin WA, around Perth and south.
Size 4–5 m high, to about 4 m wide.
Soil Grows naturally in well-drained sandy soils but adapts to crumbly loams.

Hakea plurinervia

Hakea prostrata

Climate Temperate to drier semi-tropical; frost resistant.
Aspect Best in open sunny area, unstaked to develop natural low branching and self-bracing character.
Water Moderate only; accustomed to winter rather than summer rainfall.
Feeding Sparingly with slow release or organic plant food.
Pruning May be pruned back after flowering.
Propagation Seed.
Possible Problems Seems trouble free at least in drier summer areas.

Hakea salicifolia

Hakea purpurea

FAMILY PROTEACEAE

This hakea resembles *Grevillea dielsiana*, having similar divided needle-type (terete) foliage and red spidery flowers in clusters typical of many grevilleas, rather than in lax spikes like *G. dielsiana*. Also the woody fruits which distinguish hakea from grevillea are not always obvious as in many cases they form only in small numbers if at all. This showy hakea flowers during winter and spring.

Origin Qld.
Size About 1.5 m high and nearly as wide.
Soil Adapts to both sandy and heavy loams providing drainage is good.
Climate Temperate to tropical; good frost tolerance.
Aspect Preferably full sun but at least half is needed.
Water Moderate to liberal.
Feeding Sparingly with slow release or organic fertiliser.
Pruning May be pruned back moderately after flower.
Propagation Seed.
Possible Problems Seems fairly trouble free.

Hakea salicifolia

(syn. *Hakea saligna*)

Willow-leafed Hakea

FAMILY PROTEACEAE

A hakea popularly chosen when a quick growing screen is needed. An erect small tree, not as willowy as its name implies. Leaves are lance shaped, 6–10 cm long, mid- to grey-green with attractive bronze new growth. Juvenile foliage is generally broader and rounder at the tip, with some forms tending to retain this. Small clusters of relatively insignificant white flowers appear in the leaf axils during spring.

Origin East Australian coast, especially Murrumbidgee area.
Size 6–8 m high, 4–5 m wide, but with pruning may be kept to about 3 m high.
Soil Adapts to most well-drained soils.
Climate Cooler temperate to semi-tropical.
Aspect At least half sun.
Water Liberal to moderate but reasonable drought resistance.
Feeding Slow release, organic or other complete fertilisers used sparingly.
Pruning May be pruned back each spring to desired height; otherwise after about ten years it often becomes woody and straggly.
Propagation Seed.
Possible Problems Root rot or occasionally borer may reduce life span to ten to fifteen years, especially in coastal areas. Leaf miner may attack young growth.

Hakea sericea

Needle Hakea

FAMILY PROTEACEAE

An erect, open, woody shrub, usually with heavy metallic looking branches and prickly deep green needle-like leaves 5–6 cm long. Clusters of white or

Hake purpure

Hakea sericea

sometimes pink flowers form in leaf axils during winter and spring. These are followed by hard woody capsules with some resemblance to a garden snail, up to 2.5 cm across.

Origin Eastern Australia.
Size 3–4 m high, 2–3 m wide. There are some dwarf suckering forms to about 1 m high.
Soil Occurs particularly in sandstone or granite soils but adapts to most with reasonable drainage.
Climate Cooler temperate to tropical. Accepts heavy frosts.
Aspect Dappled to full sunlight.
Water Moderate to liberal providing drainage is good.
Feeding Sparingly with slow release or organic fertiliser in spring.
Pruning May be pruned moderately after flower.
Propagation Seed.
Possible Problems Scale may attack in some areas.

Hardenbergia comptoniana

Climbing Lilac

FAMILY PAPILIONACEAE

An enthusiastic twiner that makes good cover for wire fences, trellises, old stumps, etc. Bright green leaves divided into three slender tapering oblong leaflets to 6 cm long; showy in spring with numerous sprays of small lavender blue pea flowers about 7 mm across.

Origin WA.
Size Any height desired to 3 m; will spread 2–3 m; or may be controlled by tip pruning and redirecting the thin flexible stems.
Soil Adapts to a wide range of sandy or clay-loams with reasonable drainage.
Climate Temperate to tropical; reasonably frost tolerant.
Aspect Full to half sun.
Water Liberal to moderate.
Feeding Complete plant food or fowl manure in spring.
Pruning May be pruned well back after flowering if desired.
Propagation Scarified seed.
Possible Problems Except for an occasional leaf- or bud-damaging caterpillar, is usually trouble free.

Hardenbergia violacea

(syn. *Hardenbergia monophylla*)

False sarsparilla, Happy Wanderer

FAMILY PAPILIONACEAE

Vigorous twiner with oblong to heart-shaped (cordate) leaves, 7–9 cm.long, easily distinguished from the 'chewable' sarsparilla (*Smilax glyciphylla*) by the fact that the latter has three well-defined main veins in the leaf while *Hardenbergia* has only one central main vein. In late winter–early spring the vine is almost covered with tiny violet-blue pea flowers packed closely in tapering spikes 6–8 cm long. Makes a decorative groundcover as well as fence or stump cover but given the chance will also scramble over a nearby shrub.

Origin Coast and tablelands of eastern Australia and SA.
Size To 3 m high (depending on support) 2–4 m wide.
Soil At home in lighter loams but adapts to most well-drained soils.
Climate Temperate to semi-tropical; most forms with frost tolerance.
Aspect Full to dappled sunlight, preferably with suitable support.
Water Moderate to liberal but also accepts dry periods.
Feeding Complete fertiliser or fowl manure in spring.
Pruning As required or redirect wandering leads.
Propagation Scarified seed or from selected forms such as 'Happy Wanderer', from cuttings.
Possible Problems Except for keeping it within bounds, is generally trouble free.

Hardenbergia violacea (dwarf form)

Hardenbergia comptoniana

Hardenbergia violacea

(Dwarf Form, Glenfield Form)

FAMILY PAPILIONACEAE

Apart from their restrained size, the foliage of most dwarf forms is broader, about 5 cm, broad lance shaped to oblong rather than cordate. Flowers of 'Glenfield' are usually violet like the species but a pink form now generally available is soft mauve-pink with slightly larger sprays in late winter to early spring. These may be allowed to grow as bushy shrubs, small trailers or twiners or planted at about 1 m intervals as groundcovers.

Origin Mostly eastern Australia; 'Glenfield' is from south-west of Sydney.
Size To about 1 m high, and without support a similar width.
Soil Most well drained garden soils.
Climate Temperate to semi-tropical; most with frost tolerance.
Aspect Full to dappled sunlight or half shade.
Water As for species.
Feeding As for species.
Pruning Tip pruning and cutting back after flowering retains a more compact bushy form.
Propagation Cuttings.
Possible Problems As for species.

Hardenbergia violacea

Helichrysum apiculatum

Yellow Buttons

FAMILY ASTERACEAE

An attractive mat of silver and gold. Several forms, most have lance- or wedge-shaped leaves, about 6 cm long, woolly, silvery grey in colour. Stems are also silvery and carry small clusters of golden yellow flowers mainly during spring and summer. A good rock garden or groundcover subject. Sometimes sold as *Helichrysum amplexans*, a name without official standing.

Origin Throughout all States of Australia.
Size Most forms to about 25 cm tall, 50–70 cm across. A form from south-eastern Qld has stems to 60 cm tall.
Soil Adapts to most well-drained soils but very acid sandstone soil needs addition of a little lime.
Climate Most areas but some forms may not accept heavy frost.
Aspect Open sunny position.
Water Moderate or if surface dries out between watering, liberal.
Feeding Light dressings of complete or organic fertiliser.
Pruning Remove old flower stems.
Propagation Cuttings, divisions or seed.
Possible Problems Damping off in wet humid areas. Pebbly surface scree will help.

Helichrysum baxteri

FAMILY ASTERACEAE

A low growing attractively compact little plant ideal for rock gardens. Foliage is dense, slender, linear, dark glossy green above, underside and stems grey. The fine papery petalled white flowers are about 3 cm across, with initially bright yellow zoned smooth white centres, the yellow zone gradually widening as more true florets open; flowers during spring and summer.

Origin SA, Tas., Vic. and NSW.
Size 20–30 cm high, 30–50 cm spread.
Soil Most well-drained soils, and in humid areas, like many grey or downy stemmed plants, does better with a pebbly surface scree.
Climate Cool temperate to temperate; good frost tolerance.
Aspect Full to lightly broken sunlight.
Water Moderate; allow surface to dry between waterings.
Feeding Light dressing of complete plant food or fowl manure in spring and autumn.
Pruning Prune back lightly when flowers finish.
Propagation From cuttings or seed.
Possible Problems Best renewed after two or three years especially in humid climates.

Helichrysum apiculatum

Helichrysum baxteri

Helichrysum bracteatum (Annual Forms)

Helichrysum bracteatum

(Annual Forms)

Straw Flower

FAMILY ASTERACEAE

The original species of the summer-flowering annual straw flower was a leggy plant to about 1.5 m high. Overseas plant breeders concentrating on its development have made more attractively compact forms available. Similar bright green lance-shaped foliage to about 10 cm has been retained, but is more closely set, the thickish but more branched stems terminate in flowers 3–5 cm across with short crisp papery or straw-like petals in colours ranging from white through yellows, orange, reds and pinkish tones. Picked before flower fully opens, bunched, and hung upside down until dry, they usually last and retain their colour, sometimes for years.

Origin Mostly temperate eastern Australia
Size Original species 1–1.5 m high; more recent cultivars 40–50 cm high.
Soil Adaptable to most well-drained garden soils.
Climate Cooler temperate to semi-tropical; most strains have frost tolerance.
Aspect Preferably full but at least half sun.
Water Moderate to liberal.
Feeding A little complete plant food added before sowing or transplanting.
Pruning Unnecessary; usually treated as annuals.
Propagation Seed sown in spring or in warm climates autumn sowings are usually successful.

Possible Problems Like many other annuals, damping off can occur in humid areas especially if soil surface is continuously wet; the usual leaf eaters may decide to partake of their share.

Helichrysum bracteatum

(Perennial Forms)

Straw Flowers

FAMILY ASTERACEAE

As with the annual helichrysums, several forms occur. 'Dargan Hill Monarch' has greyish leaves and comparatively large flowers to 7 cm across; it grows to 50 cm high and up to twice as wide. 'Diamond Head' has green foliage, papery yellow flowers about 3 cm across, and forms a compact plant only about 20–25 cm high and 1 m wide. As these short, thick stemmed flowers have limited appeal for flower arranging, they are cut with only about 2–3 cm of stem, wired and dried.

Origin Southern Qld and northern NSW.

Helichrysum bracteatum (Perennial Forms)

Helichrysum subulifolium

Helichrysum elatum

Size See description.
Soil Most well-drained garden soils.
Climate Cool to tropical.
Aspect At least half sun.
Water Moderate to liberal but allow surface to dry between waterings.
Feeding Light dressing of complete plant food before planting out.
Pruning Remove old flower heads.
Propagation From cuttings to reproduce as variety.
Possible Problems Check for snails or other leaf eaters. Plants in some areas deteriorate due to eel worm (nematode) attack, therefore are best treated as annuals or replaced from cuttings every two to three years.

Helichrysum elatum

(syn. *Helichrysm albicans*)

White Paper Daisy

FAMILY ASTERACEAE

A tall branching perennial with silvery white stems and lance-shaped leaves to 12 cm long which are downy beneath. The general silvery appearance is accentuated by the satiny white flowers from 2-4 cm across, the largest and earliest at the apex of plant. These appear in spring, sometimes continuing into summer.

Origin Eastern Australia, mainly along Dividing Range and foothills.
Size 1-2 m high, 1-1.5 m wide; the higher altitude forms are usually the larger.
Soil Prefers well-composted sand or crumbly loam with good drainage.
Climate Cool and temperate.
Aspect Occurs more frequently in dappled sunlight.
Water Moderate to liberal.
Fertiliser Light dressing of complete fertiliser or fowl manure.
Pruning Pruning back to about half after flowering helps delay less attractive woody growth.
Propagation Seed or cuttings.
Possible Problems in humid areas especially lower foliage browns and plant becomes unattractive after first year or may die out. Prune as described or otherwise treat as annual.

Helichrysum subulifolium

Showy Everlasting

FAMILY ASTERACEAE

A showy annual worth growing to add or prolong colour during spring and early summer after acacias, most boronias and other late winter-early spring flowers have finished; it also retains its vibrant yellow color as a dried flower. An erect plant with wiry stems and smooth, bright green, erect, but gracefully twisting, awl-shaped leaves to 12 cm long. The bright yellow papery petalled flower heads are 2-4 cm across, the size dwindling as flowering continues into summer.

Origin WA, in areas mainly between Perth and Geraldton and extending east.
Size To 50 cm high, 20-40 cm wide, but effect is best when massed or six to eight plants are set in clumps.
Soil Adapts to most well-drained soils, especially if a little lime or dolomite is added to more acid soils prior to planting.
Climate Temperate to semi-tropical with tolerance to mild frosts only.
Aspect Open sunny and preferably warm.
Water Moderate to liberal but allow surface to dry between waterings.
Feeding Mix a basic complete fertiliser into soil prior to planting.
Pruning Pinch tips from seedlings when 10-12 cm high to encourage side stems.
Propagation Seed sown in autumn, preferably direct and later thinned to 10 cm or so apart if necessary; should germinate within ten days.
Possible Problems Avoid heavy water-retentive mulch as this may cause seedlings to damp off, especially in humid areas.

Helipterum roseum

(syn. *Acrolinium roseum*)

Everlasting Daisy

FAMILY ASTERACEAE

The improved selections of this delightful 'everlasting' annual have been grown in gardens here and overseas (more frequently the latter!) for well over a century. They were a success in Kew gardens during 1854. The erect wiry stems and foliage are soft grey-green and during spring carry papery petalled daisies from 4-6 cm across in colours ranging from deep rose through paler pinks to white with yellow or black centres. Ideal for drying, picked just before or after opening.

Helipterum roseum

Origin WA, mostly south-west but also Geraldton area.
Size 30–40 cm tall, to 30 cm wide, but best grown in clumps or drifts. In nature, both flower size and height are much less.

Soil Most well-drained crumbly loams or sandy soils, preferably with a little lime or dolomite if very acid.
Climate Cooler temperate to tropical.
Aspect At least half but preferably full sun.
Water Moderate to liberal.
Feeding Respond to addition of complete fertiliser or organic manure prior to planting.
Pruning Annuals only, but pinching out centres of seedlings encourages more side branches.
Propagation From seed sown in autumn preferably direct into garden in circles for clumps or irregular squiggly drills for drifts. Little thinning out is needed if 12–15 cm spacing is left between drills.
Possible Problems Check for slugs, snails and other leaf eaters.

Hibbertia acicularis

Prickly Guinea Flower
FAMILY DILLENIACEAE

A dwarf shrub with slender brown prostrate stems, fine bristly bright green foliage and rather lax open bright yellow flowers about 3 cm across during late winter to early summer. A good trailing plant for rock gardens.

Origin Eastern Australia, especially tableland areas.
Size To 30–50 cm high, spreading to 1.5 m.
Soil Prefers a well composted gritty soil but adapts to most well-drained loams.
Climate Cool to semi-tropical; good frost tolerance.
Aspect Full to broken sunlight; the latter preferable in warm areas.
Water Moderate to liberal; drought tolerance when established.
Feeding Organic or slow release plant food in spring and autumn.
Pruning May be lightly pruned after flowering.
Propagation Cuttings.
Possible Problems Usually trouble free.

Hibbertia emptrifolia

(syn. *Hibbertia astrotricha, H. billardiera*)
Trailing Guinea Flower
FAMILY DILLENIACEAE

A showy plant with slender semi-trailing stems and small blunt elliptical leaves barely 1 cm long; generously massed with deep yellow flowers to 1.5 cm across, occurring in leaf axils, mainly

Hibbertia acicularis

during spring. It is a good plant for lightly shaded areas.

Origin All eastern States including Tas. and SA.
Size To 1 m high or leaning on other plants to 1.5 m, to 1.5 m wide, a little smaller in full sun.
Soil Prefers a moist but fairly well-drained composted sand or light loam.
Climate Cooler temperate to semi-tropical; frost tolerant.
Aspect Best in dappled or lightly broken sunlight but accepts full sun.
Water Moderate to liberal but has some drought tolerance when established.
Feeding Slow release or organic food in spring and autumn.
Pruning For more compact growth, prune back by at least one-third as flowers finish.
Propagation Cuttings.
Possible Problems Usually trouble free.

Hibbertia miniata

Copper Guinea Flower

FAMILY DILLENIACEAE

One of the rarer and more striking guinea flowers. The wiry brown stems carry rigid, dull to pale green, lance-shaped leaves to 6 cm long and slightly cupped coppery orange flowers 3–4 cm across with dark reddish bronze centre stamens, often surrounded by a broad reddish brown zone which fades as the flower ages. These appear almost throughout the year.

Origin WA.
Size To about 1 m high and as wide.
Soil Well-drained composted gritty sand or light loam.
Climate Temperate to semi-tropical.
Aspect Full to lightly broken sunlight.
Water Moderate.
Feeding Slow release or organic fertilisers applied sparingly during spring or autumn.
Pruning May be pruned moderately between flowering flushes, usually in late summer–early autumn.
Propagation Cuttings.
Possible Problems At the time of writing this is still a very difficult plant to grow successfully in cultivation, mainly because of susceptibility to root rot. Perhaps limestone chip or grafting on to resistant types such as *Hibbertia stricta* may be the answer.

Hibbertia pedunculata

FAMILY DILLENIACEAE

Small prostrate shrub with narrow dark heath-like foliage to 1 cm long and liberally decked with yellow flowers about 1.5 cm across mounted on comparatively long peduncles to stand above the foliage. These occur mainly during spring then spasmodically throughout the year. The contour-following stems often layer at the nodes.

Hibbertia pedunculata

Hibbertia emptrifolia

Hibbertia miniata

Origin Vic. and NSW.
Size 10–20 cm high, 50–75 cm wide.
Soil Prefers light composted soil with reasonable drainage.
Climate Temperate.
Aspect Lightly broken sunlight to half shade.
Water Moderate to liberal but develops drought tolerance.
Feeding Sparingly with slow release or organic fertiliser.
Pruning Rarely needed.
Propagation Cuttings or layers.
Possible Problems Usually trouble free.

Hibbertia obtusifolia

FAMILY DILLENIACEAE

An appealing little container or rock garden plant, preferably for lightly shaded areas. The small oval to wedge-shaped leaves are a glossy bright green and are liberally spangled with light yellow flowers 2–2.5 cm across, mainly in spring–early summer.

Origin Eastern Australia, Qld to Tas.
Size To about 12 cm high, but some forms to 50 cm; spreads to 50 cm.
Soil Composted sand to medium loam with good drainage.
Climate Semi-tropical to cool; good frost tolerance.
Aspect Prefers lightly dappled or broken sunlight.
Water Liberal but accepts periods of dryness when established.
Feeding Slow release or organic fertiliser sparingly in spring.
Pruning Rarely needed but cut away any thin straggly foliage.
Propagation Cuttings.
Possible Problems Fungus or perhaps eel worm sometimes browns lower foliage in humid areas. Planting on pebbly scree to stop soil splashing over stems seems to minimise this.

Hibbertia procumbens

FAMILY DILLENIACEAE

Another prostrate hibbertia forming a mat of lateral branches and dark green slender and tapering flattened leaves 1–1.5 cm long; profusely spangled with deep yellow flowers about 2.5 cm across during summer.

Origin Tas. and Vic.
Size To about 30 cm high, spreads to 1 m.
Soil Well-composted moist but well-drained peaty sand or loam is preferable but the plant is fairly adaptable.
Climate Cooler temperate to temperate.
Aspect Full to dappled sunlight, preferably the latter in warmer temperate areas.
Water Liberal to moderate but some drought resistance when established.
Feeding Slow release or organic fertiliser in spring.
Pruning When necessary for preferred shape, after flowering.
Propagation Cuttings.
Possible Problems Usually trouble free.

Hibbertia saligna

Tall Guinea Flower

FAMILY DILLENIACEAE

A medium sized erect shrub with slender, grey-green, lance-shaped, stem-clasping leaves 4–6 cm long and bright yellow saucer-shaped flowers to about 2 cm scattered along the branches, mainly during spring and early summer.

Origin NSW, coastal gullies and Dividing Range foothills.
Size To 2 m high, 1.5 m wide.
Soil Prefers damp but fairly well-drained sandy or light loamy soil with plenty of compost or well-decomposed leafmould.
Climate Cooler temperate to semi-tropical.
Aspect Usually found in damp sheltered and shaded places but adapts to dappled sunlight under trees with moderate exposure.
Water Liberal to moderate; withstands dry periods.
Feeding Sparingly with slow release or organic food.
Pruning May be pruned back by about

Hibbertia procumbens

Hibbertia obtusifolia

Hibbertia saligna

Hibbertia scandens

two-thirds as flowers finish.
Propagation Cuttings.
Possible Problems Usually trouble free.

Hibbertia scandens

Snake Vine,
Climbing Guinea Flower
FAMILY DILLENIACEAE

A vigorous and adaptable climber or trailing groundcover with thick textured ovate to lance-shaped bright green leaves 6–8 cm long and large golden yellow saucer-shaped flowers to 6 cm across appearing mainly during late spring and summer. These are followed by large calyces with five petal-like pointed lobes surrounding a wide centre displaying five large orange-red seeds.

Origin Qld and NSW, from coastal sand dunes to sheltered mountain gullies.
Size Trailing or climbing to about 2 m.
Soil Preference for well-composted light soil but also found in almost pure sand or heavy loam with reasonable drainage.
Climate Tropical to temperate; tolerates only light frost.
Aspect Full sun to light shade. In light shade it makes a comparatively light and graceful non-smothering cover for tree trunks, etc.
Water Moderate to liberal; withstands dryness when established.
Feeding Accepts moderate dressings of complete, organic or slow release fertiliser.
Pruning Repeatedly tip prune to thicken for groundcover or to create denser branching.
Propagation Cuttings, layers or seed which can be slow to germinate.
Possible Problems Usually trouble free.

Hibbertia stellaris

FAMILY DILLENIACEAE

A different little guinea flower delightful for rock gardens, containers, etc., with copper-brown wavy stems and very slender linear leaves from 3–5 cm, usually in threes; initially coppery coloured then bright green. During most of the year and particularly in spring it carries a mass of saucer-shaped coppery orange to apricot flowers about 2 cm across, sometimes lightly veiled by the wispy new coppery foliage. A wonderful companion for blue dampiera, leschenaultia, etc.

Origin Southern WA, north of Albany, often not far from the edge of swamps subject to occasional flooding.
Size 20–25 cm high and spreading 25 cm, or under favourable conditions to 40 cm.
Soil Ideal is a gritty, peaty or well-composted sandy soil with fairly good drainage and if acid a little garden lime or limestone chip added.
Climate Cooler temperate to semi-tropical; withstands all but heaviest frosts.
Aspect Lightly broken sunlight or full morning sun.
Water Moderate to liberal if soil drains well.
Feeding Sparingly with slow release or organic fertiliser.
Pruning Rarely needed but tip pruning encourages denser growth.

Hibbertia stellaris

Propagation Cuttings.
Possible Problems Sometimes short lived in hot humid summer areas due to root rot but worth growing if only for one season's flowering.

Hibbertia stricta

FAMILY DILLENIACEAE

A small erect shrub with stiff grey-green slender but blunt tipped stemless leaves to a little over 1 cm long; there are several forms with some variation in shape. Buttercup-type flowers to about 3 cm across appear mainly in spring.

Origin Temperate areas of WA, SA, Vic. and NSW.
Size To about 1 m high and nearly as wide with some size variation in the different forms.
Soil Adapts to most soils with reasonable drainage.
Climate Temperate; good frost tolerance.
Aspect Dappled to full sunlight.
Water Accepts liberal watering but once established also has good resistance to dry periods.
Feeding Slow release or organic fertiliser in spring.
Pruning May be pruned back to about half when flowers finish.
Propagation Cuttings or seed.
Possible Problems Usually trouble free.

Hibbertia stricta

Hibiscus heterophyllus

Native Rosella
FAMILY MALVACEAE

A tall shrub to small tree with thorn-studded trunk, branches and even the leaf veins; leaves usually with three deep lobes to about 12 cm long and margins serrated; upper foliage often broad lance shaped. Large yellow white or pink flowers about 10 cm or sometimes 12 cm across during summer, normally remain open for only a few hours. As the common name suggests, flower buds are edible.

Origin NSW and Qld.
Size 4–6 m tall and about 2 m wide; may be kept more compact by heavy cutting and tip pruning.
Soil Adapts to a wide range of well-drained soils, accepting strong acidity or mild alkalinity.
Climate Temperate and tropical; resents heavy frost.
Aspect Lightly broken to full sunlight.
Water Liberal but once established withstands prolonged dryness.
Feeding Needs little but accepts most complete types in small doses.
Pruning May be pruned heavily in late winter or early spring, followed by tip pruning.
Propagation Seed.
Possible Problems Its thorniness makes handling difficult.

Hibiscus huegelii

See *Alyogne huegelii.*

Hibiscus heterophyllus

Hibiscus splendens

Hibiscus splendens

FAMILY MALVACEAE

A large erect shrub with sometimes rounded but more usually three- to five-lobed leaves, 12–14 cm long. From early spring into summer, large pink flowers appear to 12 cm across, with lighter veins; they are often darker on outer side of petals.

Origin Qld and NSW.
Size 2–3 m high, about 2 m wide.
Soil Adapts to a wide range of well-drained soils.
Climate Tropical to temperate; not heavy frost.
Aspect Lightly broken to full sunlight.
Water Liberal but withstands dry periods.
Feeding Little needed but accepts animal manure, or light dressing of complete fertiliser in spring.
Pruning Prune back in winter for more compact growth.
Propagation Seed or cuttings.
Possible Problems Apart from occasional weevils in flowers seems trouble free.

Hibiscus tiliaceus

FAMILY MALVACEAE

Small usually rounded tree with large mid-green slightly downy leaves, roughly heart shaped to about 12 cm across. Large yellow flowers to 10 cm across with a dark velvety bronze-red centre appear mainly during late spring and summer.

Origin North and east of Australia, particularly near beaches and waterways, and most Pacific islands.
Size 5–7 m high, to 5 m wide.
Soil Prefers mainly sandy areas but also adapts to most loams.
Climate Tropical to warmer temperate; withstands salt spray but not frost.
Aspect At least half but preferably full sun.
Water Liberal; even accepts inundation but is also geared to survive the dry season.
Feeding Rarely needed but slow release or complete fertiliser should speed growth.
Pruning May be pruned back during early spring or in tropical areas in winter.
Propagation Seed or cuttings.
Possible Problems Usually trouble free.

Homoranthus papillatus

FAMILY MYRTACEAE

A small shrub with dense horizontal layers of ground-carpeting branches. The small leaves are grey-green and flattened. The small rather nondescript but nectar-laden flowers are cream and dull red, appearing during spring and summer. *Homoranthus flavescens*, often confused with this shrub, has coarser foliage but slightly brighter flowers. See illustration.

Origin NSW and Qld.
Size To about 30 cm high, 1 m wide.
Soil Adapts to most well-drained soils.
Climate Cool to temperate; good frost tolerance.
Aspect Needs at least two-thirds sun.
Water Liberal to moderate; accepts dry periods.
Feeding Sparing use of slow release or organic fertiliser.
Pruning Only if needed for shaping.
Propagation Cuttings.
Possible Problems With ample sunlight and drainage, should be trouble free.

Hovea longifolia

FAMILY PAPILIONACEAE

An erect and fairly sparse shrub with slender oblong leaves to 7 cm long, dark green above, greyish beneath. Small mauve-blue to purple pea flowers are borne singly or in short racemes during late winter and spring or in cool temperate areas into early summer.

Origin All States except WA and NT.
Size 1–2 m, to 1 m wide.

Homoranthus papillatus

Hibiscus tiliaceus

Hovea longifolia

Soil Adapts to most well-drained soils.
Climate Cool temperate to semi-tropical; good frost tolerance.
Aspect Lightly broken sunlight to light shade.
Water Moderate to liberal; tolerates dry periods.
Feeding Slow release or sparingly with

complete fertiliser as new spring growth starts.
Pruning Prune back by about one-third after flowering.
Propagation Scarified seed.
Possible Problems Usually trouble free.

Hybanthus filiformis

Hybanthus filiformis

(syn. *Oinidium filiforme*)
Lady's Slipper
FAMILY VIOLACEAE

A small plant, often found straggling in native grassland or other small shrubby growth. It has 5-6 cm long slender linear leaves, lower ones often broader and lance shaped, scattered sparsely in pairs along wiry erect stems, terminating in short loose spikes of flowers with one prominent lower petal. Flowers are bluish mauve to pale violet with a deeper zone around a slender white base. The other four petals are so minute that they appear absent. Flowers mainly during late winter and spring.

Origin Eastern mainland States.
Size From 20-50 cm high and often as wide.
Soil Light sandy and well-composted to medium loam, preferably fairly moist but well drained.
Climate Temperate to semi-tropical; will not tolerate severe frost.
Aspect Full to lightly broken sunlight.
Water Liberal to moderate.

Hymenosporum flavum

Feeding Light scattering of complete plant food or slow release fertiliser in autumn and early spring.
Pruning Not necessary.
Propagation Seed or cuttings.
Possible Problems Usually trouble free.

Hymenosporum flavum

Native Frangipani
FAMILY PITTOSPORACEAE

A quick growing and attractive slender tree with glossy dark green slender elliptical (correctly oblanceolate) leaves to about 15 cm long. During spring carries clusters of fragrant flowers with some resemblance to round petalled frangipani. They open limey cream and age to mustard yellow. Characteristically, the branches occur in well-separated tiers of gradually diminishing height, pagoda style.

Origin Qld and NSW rainforest country.
Size Average height to 10 m with a spread about head height to 4-5 m; larger in native habitats and more stunted in exposed windy areas.
Soil Well drained, all but heavy puggy clay or very alkaline.
Climate Tropical to cooler temperate. Reasonable frost tolerance when established. Grows with some cold wind shelter in Hobart and Canberra.
Aspect May be started in light shade to eventually reach sunlight (remains leggy until it does). Some shelter from cold winds desirable.
Water Liberal but once established has some drought resistance.
Feeding Responds to water soluble fertilisers to speed early growth; also slow release, complete fertiliser or animal manure.
Pruning Out of character with tree and does little to encourage bushiness.
Propagation Seed.
Possible Problems Usually trouble free.

Hypocalymma angustifolium

FAMILY MYRTACEAE

A fairly open and erect but attractively rounded little shrub with very slender liniear leaves, 3-4 cm long, and from late winter into mid-spring the usually straight slender branches are clustered with fluffy white or pink flowers.

Origin Southern WA.
Size About 1m high, to 1.5 m wide.
Soil Originates in peaty but gritty sand kept fairly moist by nearby swamps but accepts average well-drained garden loam or sandy soils with compost added.
Climate Cooler temperate to semi-tropical.
Aspect Lightly broken to full sunlight with some preference for the former.
Water Liberal but also accepts periods of dryness.
Fertiliser Slow release organic or light dressings of complete fertiliser.
Pruning Improved by pruning back by about one-third after flowering.
Propagation Cuttings
Possible Problems Usually trouble free.

Hypocalymma cordifolium

'Golden Veil'

FAMILY MYRTACEAE

A lovely compact but soft looking foliage plant for accent, colour or complementary contrast especially combined with grey or dark green shrubbery. The species *Hypocalymma cordifolium* has its coppery red stems clustered with deep green, broad, heart-shaped leaves 1.5–2 cm long. Those of 'Golden Veil' are heavily variegated with soft cream and the stem colour seems to diffuse into the new tip growth giving a pinkish flush. Small creamy white flowers appear in leaf axils during spring.

Origin Species is from WA; 'Golden Veil' is from a Victorian nursery.
Size 1m high, to about 1.5 m wide.
Climate: Cool temperate to semi-tropical.
Aspect At least half sun for good foliage colour.
Water Moderate to liberal; dryness during hot weather can result in some foliage burn.
Feeding Slow release or organic fertiliser.
Pruning Tip prune or cut back lightly before new spring growth starts for more compact growth.
Propagation Cuttings
Possible Problems Foliage burn can occur through liberal use of complete fertiliser, especially if dryness follows application.

Hypocalymma angustifolium

Hypocalymma cordifolium 'Golden Veil'

Indigofera australis

Australian Indigo

FAMILY PAPILIONACEAE

This shrub has a fern-like appearance created by the gracefully arching branches massed with soft blue-green pinnate foliage divided into small elliptical leaflets. During spring and often into summer, numerous erect sprays of tiny rose pea flowers sit above the branches like groups of tapering candles.

Origin All States of Australia both coast and inland.
Size 1.5–2 m high, to 2 m across.
Soil Any well-drained soil.
Climate Cool to tropical. Frost tolerant, even in Scotland's Botanic Gardens.
Aspect Flowers well in either full sun or light shade.
Water Moderate or liberal; good drought resistance when established.
Feeding A lower fertility plant but accepts standard complete fertilisers applied moderately.
Pruning More compact and denser foliage if pruned back after flowering.
Propagation Scarified seed.
Possible Problems May be visited by an occasional leaf-rolling caterpillar; not of serious consequence.

Ipomea brasiliensis

(syn. *Ipomea pes-caprae*

Goat's Foot Convolvulus

FAMILY CONVOLVULACEAE

A common plant near the shoreline of the warmer temperate and tropical Australian coastline and particularly the Pacific Islands. It has unusual large smooth, stiff textured leaves cleft at the apex to give a goat's hoof appearance. The showy pink- and purple-centred flowers, 6–8 cm across appear throughout the year.

Origin As above.
Size To about 15 cm high, trialing 2–4 m.
Soil Usually found in sand but also adapts to heavier loam.
Climate Frost free temperate and tropical.
Aspect Full to lightly broken sunlight.

Indigofera australis

Isopogon anemonifolius

Ipomea brasiliensis

Water Liberal but can withstand dry periods.
Fertiliser Complete or slow release if needed.
Pruning Rarely applicable.
Propagation Seed.
Possible Problems Usually trouble free.

Isopogon anemonifolius

Drumsticks
FAMILY PROTEACEAE

Small spreading shrub, worth growing for its decorative soft light green foliage. Individual fan-shaped leaves about 10 cm long are flattened and several times divided into segments only about 1 mm wide, giving a lacy appearance, and during spring frame the terminal yellow flower heads which are 3–4.5 cm across and later the mature silver-grey globular seed cones.

Origin Vic., NSW and Qld.
Size 1–1.5 m tall and wide.
Soil Most well-drained soils, preferably with a leafy mulch.
Climate Cool to semi-tropical; good frost and sea spray tolerance.
Aspect Full to lightly broken sunlight.
Water Liberal or moderate; good drought resistance.
Feeding Sparingly, with slow release or organic; a leafy mulch is really all that they ask.
Pruning Stems may be pruned down to about 10 cm to encourage base branching; better for general effect to prune only about one-third of the stems each year.
Propagation Seed which normally germinates within a few weeks.
Possible Problems Usually trouble free.

Isopogon anethifolius

Drumsticks
FAMILY PROTEACEAE

Resembling *Isopogon anemonifolius* but is even softer looking as the longer leaves (to 16 cm) are divided into fine needle-like (terete) segments. Young growth often begins a soft copper colour then merges through reddish bronze to green. The yellow flower heads are a little longer than *I. anemonifolius* and continue into early summer; the globular seed cones (drumsticks) are larger, to 2.5 cm across.

Origin NSW.
Size 1–1.75 m high, to 1.5 m across; much lower in exposed coastal areas.
Soil More often in light sandstone-type soils but adapts to heavier well-drained loams.
Climate Cooler temperate to semi-tropical; frost tolerant.

Aspect, water, feeding, pruning and possible problems as for *I. anemonifolius.*

Isopogon divergens

Spreading Cone Bush

FAMILY PROTEACEAE

This showy isopogon has erect growth but makes a dense spreading clump of stems. Leaves are finely divided into needle-like segments and carry heads of soft woolly looking silvery rose pink blooms in spring.

Origin WA, Darling Range area.
Size To 1.75 m high and eventually to 2 m across.
Soil Naturally occurring in gravelly soil, adapts to well-drained gritty composted and slightly limey loam.
Climate Temperate.
Aspect Best in just lightly broken sunlight but also occurs naturally in full sun.
Water Moderate only; resents wet summers.
Feeding Sparingly if at all with slow release or organic food.
Pruning Stems may be pruned well back after flowering.
Propagation Seed.
Possible Problems Difficult to grow on east coast. Well-drained soil with limestone chip added may help.

Isopogon divergens

Isotoma axillaris

Isotoma axillaris

FAMILY LOBELIACEAE

A compact and attractively clumpy plant with narrow lobed light green leaves to about 8 cm long. During late spring and summer it gives a fine display of erect stemmed starry blue to pale violet flowers, 2–3 cm across. Excellent for rock gardens or containers.

Origin Vic., NSW and Qld.
Size 25–30 cm high, 40–50 cm wide.
Soil Adapts to most well-drained soils, preferably with plenty of compost added.

Isopogon anethifolius

Climate Temperate to semi-tropical, without great frost tolerance, but it is quick to flower and may be treated as an annual in very cold winter districts.

Aspect Full to lightly broken sunlight.
Water Liberal, especially during spring and summer but let surface soil dry between waterings.
Feeding Responds to light dressings of complete or slow release plant food when planting out and/or in spring.
Pruning Prune back old growth when flowers finish.
Propagation Seed — in warmer areas sown in autumn or in frosty districts about July under shelter.
Possible Problems Avoid contact with sap as it is an eye irritant and may also affect skin.

Kangaroo Paw

See *Anigozanthus*

Kunzea ambigua

FAMILY MYRTACEAE

An erect shrub with fine slightly greyish to olive green foliage, usually dense on the gracefully arching outer branches which are most decorative when clustered with fluffy white flowers in late spring to early summer. In fairly open situations with occasional pruning it can make a good, reasonably dense screen. There is also a pink form, usually a little more sparse in growth and flower.-Lasts well as a cut flower.

Origin Tas, Vic. and NSW.
Size About 2-2.5 m high and as wide in sunny situations but nearly twice as tall in lightly timbered areas.
Soil Grows well in acid sandstone or in slightly clayey or shale areas with reasonable drainage.
Climate Cool to semi-tropical; good frost tolerance.
Aspect Full to dappled sunlight; usually taller and higher branching in latter.
Water Liberal to moderate; good drought resistance when established.
Feeding Most complete type fertilisers or fowl manure used in moderation.
Pruning For more compact growth prune fairly hard as flowers finish.
Propagation Cuttings.
Possible Problems Usually trouble free.

Kunzea baxterii

FAMILY MYRTACEAE

A showy shrub with closely set, slender, lance-shaped to oblong leaves about 2 cm long and brilliant scarlet to blood red bottlebrush-type flowers from mid-winter into mid spring.

Origin WA, south-west to southern coast — Stirling Ranges and east.
Size To 2 m high and as wide.
Soil Adapts to a wide range of slightly acid soils, from the rocky and gritty sands of its natural habitat to medium loams with good drainage and preferably a leafy mulch.
Climate Cooler temperate to semi-tropical.
Aspect Lightly broken to full sunlight.
Water Liberal to moderate.
Pruning Prune back as flowers finish in late spring for more compact growth.
Propagation Cuttings or seed; from seed it may take five years to flower.
Possible Problems Usually trouble free except perhaps for root rot in hot wet areas; can be grafted on to more tolerant *Kungea ambigua*.

Kunzea parvifolia

FAMILY MYRTACEAE

This shrub appears sparse and twiggy because of its insignificant heath-like foliage but it is most appealing during spring and into summer when it is generously decorated in fluffy mauve-pink balls of flower.

Origin Vic. and NSW, including ACT.
Size About 1.5 m high and spreading 2.5-3 m.
Soil Adapts to most garden soils with moderate drainage.
Climate Cooler temperate to semi-tropical.

Kunzea ambigua

Kunzea ambigua (pink form)

Kunzea baxterii

Kunzea parvifolia

Aspect Full to lightly dappled sunlight.
Water Liberal to moderate but withstands dry periods.
Feeding Very sparingly with slow release or organic fertiliser.
Pruning Prune back as flower finishes to keep compact.
Propagation Seed or cuttings.
Possible Problems Usually trouble free.

Lachnostachys eriobotyra

Lamb's Tails

FAMILY VERBENACEAE

A most unusual and arresting large branching shrub with woolly silver-grey linear leaves to 8 cm long and dense white woolly spikes of a similar length studded with small purple-centred flowers. These spikes are in dense showy clusters at the ends of the branches and sometimes also in an unusual cross-formation formed by four long spikes on opposing branchlets.

Origin WA, particularly on the warmer

Lachnostachys eriobotyra

Lagunaria patersonii

Leschenaultia biloba

sandplains south of Geraldton.
Size Usually about 1.5 m but occasionally larger and as wide.
Soil At home in warm gritty slightly limey sand with good drainage. At time of writing there is no record of successful adaptability to other soils.
Climate Temperate to drier semi-tropical.
Aspect Open sunny position with lower groundcovering plants.
Water Moderate only.
Feeding Suggest light application of complete organic fertiliser.
Pruning Should respond to light pruning after flowering.
Propagation Seed, but difficult to obtain viable material from inflorescences; also from cuttings, but not in conventional humid situations.
Possible Problems Seems unsuitable in all but a warm, moderately dry climate, possibly with slightly alkaline soil.

Lagunaria patersonii

Norfolk Island Hibiscus

FAMILY MALVACEAE

An adaptable, usually erect pyramid-shaped tree with leathery elliptical leaves to 10 cm long, dark green above, grey underneath and white to soft pink waxy bell-shaped flowers opening at the lip, to about 5 cm across, mainly from late spring through summer. Salt spray resistant and excellent for the beachfront, except that the seed capsules contain hair-like spines that can penetrate skin and cause irritation.

Origin Norfolk Island and Qld.
Size Usually to about 10 m and 4–5 m wide, but old naturals on Norfolk Island have trunks to about 1.5 m diameter and a spreading top to about 10 m or more.
Soil Adapts to almost pure sand but seems happier with at least a medium loam. Good drainage needed.
Climate Temperate to tropical with tolerance to moderate frosts.
Aspect Best in open sunny position but grows and flowers in light shade.
Water Liberal to moderate.
Fertiliser Accepts both standard complete and organic fertiliser, used moderately.
Pruning Not normally needed but permissible in spring.
Propagation Seed.

Lambertia formosa

Possible Problems Usually trouble free, except for the limitations above.

Lambertia formosa

Mountain Devil

FAMILY PROTEACEAE

Usually a fan-shaped fairly open shrub with glossy sharp pointed linear leaves to 5 cm long and erect clusters of tubular florets 3–4 cm long with prominently protruding straight styles. The flower is enclosed in slender transluscent reddish to lime bracts longer than the florets. These flowers give rise to a quaint woody seed capsule with a warty 'face', two erect 'horns' and a long upturned 'nose'. They are sometimes attached to pipe cleaners and offered to tourists as 'Mountain Devils'. Flowering time mostly late winter–spring but also sporadic.

Origin NSW, mostly coast and foothills. Other species come from WA but these mostly need limestone soil to succeed.
Size 1–2.5 m high and nearly as wide, or in exposed coastal and highland areas as low as 60 cm.
Soil More widely in sandstone but also adapts to moderately heavy loam with reasonable drainage.
Climate Cooler temperate to semi-tropical.
Aspect Full sun to light or part shade.
Water Liberal but accepts dry conditions when established.
Feeding Sparingly if at all with slow release or organic plant food.
Pruning May be shortened back by about one-third after main flowering, or alternate stems shortened much lower each year.
Propagation Seed.
Possible Problems Usually trouble free but in nature where subjected to an occasional bushfire they seem more vigorous and floriferous.

Leschenaultia biloba

Blue Leschenaultia,
Mirror of Heaven

FAMILY GOODENIACEAE

A small usually spreading shrub with fine bluish green foliage to about 1 cm long clustering the slender arching branches. It is a glorious sight during mid-winter to late spring when massed with flowers about 2 cm across, bluer than the bluest sky. However there are colour variations, some a light icy to sky blue, others a dark royal blue, perhaps with a pure white centre blaze. They have a reputation for short life, especially in the humid east coast areas, but they flower within a year from cuttings or a few months longer from seed, therefore are well worth growing, even for the pleasure of only one seasons' beauty.

Origin WA, mainly the south-west but extending through the Darling Range and north of Perth, usually lighter in colour as they extend northwards.
Size About 0.5–0.75 m high and usually a little wider.

Leschenaultia formosa

Soil Ideal in elevated pocket with composted gritty sand or gravelly soil, preferably with a pebbly scree but with some low cover (ideally hibbertias, tetrathecas or native grasses) to keep soil cool.
Climate Temperate.
Aspect Full to lightly broken sunlight.
Water Liberal to moderate in winter then moderate (allow soil to dry about 2 cm deep between waterings).
Fertiliser Slow release just sparingly or a little compost.
Pruning Pruning back one-third to one-half after flowering usually prolongs life.
Propagation Cuttings preferably, or seed.
Possible Problems Loss due to overwet or excessively hot soils, especially in humid coastal area; normally has a fairly short life.

Leschenaultia superba

Leschenaultia formosa

Red Leschenaultia

FAMILY GOODENIACEAE

A delightful little plant for rock garden pockets, containers or for grouping in the foreground of taller shrubbery and outstanding in company with the prostrate hibbertias, dampieras, etc. It makes a dense cushion of fine branches packed with tiny green heath-like foliage. From late autumn to late spring it is almost covered with flowers over 1 cm across, more usually bright pillarbox red but it also occurs in yellow, scarlet pinkish tones and magenta.

Origin WA, southern area between Albany and Esperance.
Size 15–18 cm high, 30–45 cm wide.
Soil Occurs naturally in peaty sand but adapts to most well-drained loams if lightened with coarse sand and compost.
Climate Cooler temperate to temperate; most forms with tolerance to at least moderate frosts.
Aspect Best with lightly broken sunlight through the hottest part of the day.
Water At least moderate but not too frequently wet during summer, especially in humid districts.
Feeding No more than a light sprinkling of slow release or a soil dusting with organic fertiliser.
Pruning An advantage to clip off old flower heads.
Propagation Seed or cuttings.
Possible Problems Root fungi in hot wet situations.

Leschenaultia superba

FAMILY GOODENIACEAE

A prostrate little shrub with attractive dense blue-green needle-like foliage about 2.5 cm long some stems ground carpeting, others ascending. Long orange red tubular flowers, 2–3 cm across, with three prominent lobes at the tip, occur in twos and threes spasmodically from winter through into summer.

Origin Southern WA (Esperance area).
Size To 35–40 cm high, to about 75 cm wide.
Soil Gravelly sand is natural habitat but adapts to most well-drained soils, either with a pebbly scree or leafy surface mulch.
Climate Temperate to cool; withstands moderate frost.
Aspect Best with lightly broken sunlight or at least some protection during the hottest part of day.
Water Liberal to moderate; avoid frequently wet surface soil in humid summer districts.
Feeding Very sparingly if at all with slow release or organic fertiliser; a little compost should be sufficient.
Pruning Prune back ascending growths occasionally for more compact growth.
Propagation Cuttings or seed; seedlings should flower in the second year.
Possible Problems Leaftip-rolling caterpillars may attack but have little more than a tip-pruning effect.

Leptospermum flavescens

FAMILY MYRTACEAE

The species is a compact medium to tall shrub with slender pale green leaves to about 2 cm. During late spring and early summer the outer branchlets are pendulous with the weight of white to creamy white flowers about 2 cm across. Variety 'Pacific Beauty' is semi-prostrate and cascading with dense foamy looking sprays of white flowers, ideal for rockeries or cascading down embankments.

Origin Qld and NSW.
Size The species grows to 3 m high and 2–2.5 m across. Variety 'Pacific Beauty' grows to 60–70 cm high and 1.5–2 m across.
Soil Adapts to most well-drained garden soils.
Climate Tropical to temperate with at least moderate frost tolerance.
Aspect Best in full sun but accepts dappled sunlight; good salt wind tolerance.
Water Liberal to moderate; reasonable tolerance to dryness when established.
Feeding Accepts light dressings of slow release, organic or other complete fertilisers, preferably in early summer and watered well afterwards.
Pruning May be pruned back as flowers finish but at least a few leaves should be allowed to remain on each stem.
Propagation Species seed or cuttings; 'Pacific Beauty' from cuttings only.
Possible Problems May be attacked by leaf-webbing caterpillar.

Leptospermum flavescens var. 'Pink Cascade'

FAMILY MYRTACEAE

A beautiful low growing to prostrate shrub with foliage similar to that of the

Leptospermum flavescens

Leptospermum flavescens var. 'Pink Cascade'

species, similar size flowers packed in slender sprays in spring, not as heavily pendulous but probably in greater quantity and spreading horizontally as well as downward. Colour is a delightful shade of soft pink.

Origin Cultivar from *Leptospermum flavescens.*
Size 50–60 cm high and spreading 75 cm and possibly to 1 m.
Soil Adapts to most well-drained garden soils with some preference for composted sandy loams, in all cases with a leafy surface mulch.
Climate Tropical to cooler temperate areas.
Aspect At least half sun.
Water Liberal to moderate.
Feeding Light application of slow release or organic fertiliser.
Pruning Prune back moderately after flowering.
Propagation Cuttings only.
Possible Problems As species.

Leptospermum laevigatum

Coastal Tea Tree

FAMILY MYRTACEAE

A large densely foliaged shrub with grey-green slender oval leaves to 2 cm and, during spring or early summer, with the outer branchlets clustered with white flowers to 1.5 cm across. Good salt wind tolerance and often used as windbreak or to resume sand dunes.

Origin SA, and all eastern States including Tas. especially near the coast.
Size 5–6 m and usually spreading 2–3 m.
Soil At home in sandy soils but adaptable.
Climate Tropical to cooler temperate, good frost tolerance.
Aspect Full to half sun; excellent salt wind tolerance.
Water Liberal to moderate.

Leptospermum laevigatum

Feeding When established accepts light to moderate applications of complete, slow release or organic fertilisers.
Pruning May be pruned back after flowering to encourage lower branching.
Propagation Cuttings or seed; the latter will open naturally on the shrub.
Possible Problems Except for tea tree scale which can occur under some conditions, it is reasonably trouble free.

Leptospermum persicifolium

See *Leptospermum squarosum.*

Leptospermum petersonii

Lemon Scented Tea Tree

FAMILY MYRTACEAE

An excellent fast growing light screen or large background shrub with pendulous outer branches and bright glossy green slender elliptical leaves to 4 cm long emitting a pleasantly refreshing lemon aroma when crushed. New growth is reddish bronze. Small white flowers about 1.5 cm across, with green centres, appear along the outer branches in spring.

Origin Qld and NSW.
Size 4 m high, 3–4 m wide, in some cases larger.
Soil Very adaptable. Accepts light sandy soil or fairly heavy clay but the latter should be lightened with coarse sand and/or leafmould to encourage rapid growth.
Climate Temperate to tropical; tolerates moderate frost.
Aspect Full sun to half shade.

Water Liberal to moderate; once established withstands long dry periods.
Feeding Complete soluble fertiliser or slow release applied in spring should speed growth.
Pruning May be pruned back very heavily but pruning is rarely necessary.
Propagation Seed or cuttings.
Possible Problems Relatively trouble free.

Leptospermum rotundifolium

See *Leptospermum scoparium* var. *rotundifolium*

Leptospermum scoparium

And cultivars

FAMILY MYRTACEAE

The species is a fairly dense shrub with narrow lance-shaped leaves about 1 cm long; in spring or early summer white flowers appear, about 1 cm across, occasionally pink tinged. Form varies with locality of origin, between just under 2 m to over 3 m and spreading nearly as wide.

Origin From all eastern States and New Zealand.
L. scoparium cultivars have been attractive, colourful and popular garden shrubs since the 1930s. Among the best of these was Vic. raised *L. s. Lambethii* with large light and dark pink flowers from autumn to spring, growing to about 3 m high and 2 m wide. Other popular cultivars originating from Californian selection include: *L. s.* 'Fairy Rose' with dark foliage and double pink flowers, growing about 2 m and nearly as wide; *L. s.* 'Red Damask' with deep red double blooms, growing 2.5 m high and to 1.75 m wide; *L. s.* 'Scarlet Carnival', a lighter red double and growth a little shorter than 'Red Damask'.
L. s. 'Snow Flurry' is an outstanding double white to about 2.5 m high and 1.5 m wide. Most have a life span of only about ten years but they give so much within that time.
Soil Suit most well-drained garden soils.
Climate Cool temperate to semi-tropical; good frost tolerance.
Aspect Full sun for the best flower display and compactness of growth.

Leptospermum petersonii

Leptospermum scoparium lambethii

Water Moderate to liberal but once established tolerates dry periods.
Feeding Sparingly to moderately with slow release, complete, or organic fertiliser, including fowl manure, when new growth starts after pruning.
Pruning Prune back by at least one-third as flowers finish.
Propagation Species from seed; cultivars only from cuttings.
Possible Problems Leaf stripping by webbing caterpillars. Scale which is minute but indicated by black sooty mould on foliage and stems.

Leptospermum scoparium Rosy Ruffles

Leptospermum squarosum

Leptospermum scoparium var. *rotundifolium*

Leptospermum scoparium var. *rotundifolium* (white form)

Leptospermum scoparium var. ***rotundifolium***

(syn. *Leptospermum rotundifolium*)

Round Leafed Tea Tree

FAMILY MYRTACEAE

The former official name seems more appropriate as this beautiful tea tree has little resemblance to *Leptospermum scoparium* in form, foliage or flower. The slender branches are rather horizontal and during spring and summer carry neat glossy circular foliage 6–7 mm in diameter and beautifully simple large flowers about 2.5 cm across with conspicuous centres like pale green crystal. Petal colour of different forms varies from clear soft pink to a now widely sold lavender and there is a white form that grows on NSW southern tablelands.

Origin NSW.
Size To about 2 m high and spreading to 3 m.
Soil Adapts to most well-drained garden soils.
Climate Cool to semi-tropical. Good frost tolerance.
Aspect Full to lightly broken sunlight.
Water Liberal to moderate; withstands dryness when established.
Feeding Sparingly with slow release, organic or standard complete fertiliser when new growth commences.
Pruning Prune back after flowering for more compact growth.
Propagation Cuttings to reproduce type, otherwise seed.
Possible Problems Seems less prone to webbing caterpillar and to scale than some other *L. scoparium* types.

Leptospermum squarosum

(syn. *Leptospermum persiciflorum*)

Peach Flowered Tea Tree

FAMILY MYRTACEAE

An erect large shrub, usually with a single stout trunk and fairly sparse arching to pendulous branches, stiff lance-shaped foliage to 1.5 cm long and inner branchlets liberally clustered with beautiful large white, frequently flushed or occasionally deep pink flowers 2–2.2 cm across, mainly from late autumn to spring. It seems unfortunate that these come mainly on the previous season's growth. A good tree to provide the lightly broken sunlight for boronias and other smaller shrubbery and still provide interest.

Origin Southern Qld and NSW.
Size 2.5–4 m high and about 2 m wide.
Soil Occurs mainly in sandstone country just below the ridges but adapts to most lighter, or even medium loams with good leafy mulches.
Climate Cooler temperate to semi-tropical.
Aspect Full or lightly broken sunlight.
Water Fairly liberal, occurs mainly where soil is damp with seepage but when established accepts long dry periods.
Feeding Accepts light application of slow release or organic plant foods but survives happily with only a leafy mulch.
Pruning Not usually necessary except to cut away old twiggy growth but if preferred, alternate branches may be cut back heavily in spring so that flower is not completely lost the following year.
Propagation Seed, or to retain parent character, cuttings.
Possible Problems Usually unaffected and longer lived than *Leptospermum scoparium* forms.

Libertia paniculata

FAMILY IRIDACEAE

Perennial clumps of deep green slender strap-type arching foliage and during spring erect wiry branching stems carrying a succession of oval buds opening to white flowers to about 3 cm across.

Origin Vic, NSW and Qld.
Size Foliage clumps to about 40 cm high, spreading 50–75 cm; flower stems 50–60 cm.
Soil Adapt to most garden soils.
Climate Cool to semi-tropical.
Aspect Best with dappled sunlight or full morning sun only. Foliage tends to yellow with full exposure.
Water Liberal but some resistance to dry periods.
Feeding Light dressing of complete fertiliser, slow release or animal manure during autumn and spring.
Pruning Remove old flower stems and untidy foliage.
Propagation Division of clumps in autumn or seed.
Possible Problems Usually trouble free.

Libertia paniculata

Lomandra longifolia

Lomandra longifolia

FAMILY XANTHORRHOEACEAE

Tough clumpy plant with fairly erect and stiff, but later arching, glossy green strap-like foliage to 2 cm wide terminating abruptly with a spine-like projection either side, and in spring, buff-yellow flower spikes 10–15 cm long and flattened stems to about the same length. In spite of their armament of rather prickly spines they are sometimes popular for floral decoration. The plant offers contrast of foliage and form in the garden and survives in difficult situations.

Origin Entire length of Australian east coast including Tas.
Size 50–70 cm high and gradually spreading to about 1 m.
Soil Flourishes in inhospitable shallow sandstone pockets or heavy clay-loams, preferably with reasonable drainage.
Climate Cold to tropical.
Aspect Full sun or shade; needs some sun to flower.
Water Accepts flood or drought; give enough to establish the plant then forget it.
Feeding Seems to need none but probably hard to kill with overfeeding.
Pruning Only when in the way, then use a spade.
Propagation Division (cut back foliage) or seed.
Possible Problems May get too prickly for comfort if in the wrong situation; it moves easily.

Lomatia silaifolia

Parsley Fern

FAMILY PROTEACEAE

A small shrub with suckering canes grown mainly for its decorative foliage. The leaves, about 20 cm long and 12 cm wide, are divided several times into wiry segments like a stiff piece of gracefully arching netting. During summer, erect creamy white flower spikes to 30 cm long stand above the foliage.

Origin Qld and NSW.
Size 0.75–1.5 m high and similar width.
Soil Occurs naturally in sand with leafy mulch and the heavier shale or clay-loam with good drainage.
Climate Cooler temperate to semi-tropical.
Aspect Full sun to shade.
Water Liberal but once established drought resistance is good.
Feeding Sparingly with slow release or organic fertiliser.

Lomatia silaifolia

Pruning Normally not necessary but they recover well from cutting or burning all above soil growth.
Propagation Seed.
Possible Problems Usually trouble free but tiny leaf hoppers can cause flower buds to drop.

Melaleuca armillaris

Bracelet Honey Myrtle

FAMILY MYRTACEAE

Small tree or large bushy shrub with ridged brown bark and dark green slender linear leaves to 2.5 cm long. Slender cylindrical brushes of white flowers occur in spring or early summer usually towards the centre rather than tips of the outer branchlets. A good windbreak providing it is not staked during early growth then later becomes top heavy. Stands salt spray.

Origin Vic., NSW and Qld, mainly in coastal areas.
Size 5–6 m high, to 7 m wide.
Soil Adapt to a wide range of soils from coastal sand to shale or heavy clay. Prefers reasonable drainage.
Climate Temperate to tropical; sensitive to heavy frost when young.
Aspect Best in open sunny situation.
Water Liberal to moderate.
Feeding Resents heavy feeding; slow release or organic fertiliser is safest.
Pruning Prune back heavily at planting time to encourage self-bracing low branches — and throw away the stake.
Propagation Seed or cuttings.
Possible Problems Top-heaviness; see pruning, or plant at 45° angle, facing away from the wind.

Melaleuca bracteata

Melaleuca bracteata

FAMILY MYRTACEAE

Fast growing small erect bushy tree with ridgy bark and lance-shaped leaves, 1.5–2 cm long; 5 cm long cylindrical brushes of white flowers occur in summer. A good screen tree and tolerates wet soil.

Origin Qld, NT, SA and NSW.
Size 6–7 m high, 3–5 m across.
Soil Adapts to most soils and accepts bad drainage or shallow soil over rock.
Climate Temperate to tropical; tolerates most frosts.
Aspect Best in at least half sun.
Water Liberal to moderate; although suitable for wet soils it also withstands periods of dryness.
Feeding Sparingly with slow release or preferably organic based complete fertiliser.
Pruning May be pruned back heavily if necessary.
Propagation Seed or cuttings.
Possible Problems Usually trouble free.

Melaleuca bracteata var. 'Golden Gem'

Melaleuca armillaris

Melaleuca bracteata var. 'Golden Gem'

FAMILY MYRTACEAE

A beautiful golden foliaged form of above, smaller growing, good for foliage contrast or tub specimen.
Size About 2 m high, 1.5 m wide.
Treatment as *Melaeuca bracteata*.

Melaleuca diosmifolia

Melaleuca fulgens (salmon form)

Melaleuca fulgens

Melaleuca bracteata var. 'Revolution Green'

FAMILY MYRTACEAE

This cultivar is similar to the species in foliage but is lower growing (to about 2 m high and 1.5 m wide); it is useful where a small screen is needed. Treatment as for *Melaleuca bracteata*.

Melaleuca diosmifolia

FAMILY MYRTACEAE

This shrub's outstanding feature is the arrangement of its foliage. Rather rigid green oval to elliptical leaves to about 1 cm long are tightly and evenly packed along the long slender branches making them look like long curved rough textured cylinders or a little like the Chilean Monkey Puzzle Tree. The flower brushes are a light green (to avoid detracting from the foliage) and appear mainly in spring to early summer.

Origin WA, Stirling Range area.
Size 2.5 m high, about 2 m wide.
Soil Occurs naturally in gravelly sand but adapts to most well-drained loams except heavy clay.
Climate Temperate to semi-tropical with only light frost.
Aspect Best in full sun or lightly broken sunlight.
Water Liberal to moderate.
Feeding Sparingly with slow release or organic fertilisers.
Pruning Sometimes trimmed as a hedge but removing the old flower brushes or dead branches seems to mask its character.
Propagation Seed or cuttings.
Possible Problems Seems trouble free.

Melaleuca fulgens

Scarlet Honey Myrtle
FAMILY MYRTACEAE

A small shrub with slender bluish green lance-shaped leaves to about 3 cm long, slightly concaved, grown for the brilliance of its scarlet brushes, their brightness accentuated when the stamens are tipped with golden pollen. There is also a purple form and a delightful deep apricot.

Origin WA, from east of Geraldton down to cooler summer areas near Esperance.

Size 1–1.5 m high, to 1.5 m wide.
Soil Needs good drainage, preferring the gritty sandy soils and lighter loams with a leafy mulch.
Climate Cooler temperate to semi-tropical; some frost protection needed in very cold districts.
Aspect Full to lightly broken sunlight.
Water Moderate; tolerates liberal if well drained.
Feeding Sparingly; light blood and bone, poultry manure or slow release.
Pruning May be pruned back moderately after flowering.
Possible Problems Can develop root rot, especially in hot wet summer areas but well worth perseverance.

Melaleuca hypericifolia

Melaleuca hypericifolia

Robin Red Breast Honey Myrtle

FAMILY MYRTACEAE

An adaptable and attractive dense foliaged shrub for screening, backgrounds, etc., with light to mid-green lance-shaped leaves, 4 to 5 cm long, as the species name suggest, it is like hypericum (the shrubby types). The red flower brushes occur mainly during late spring and summer on older wood, therefore are partly hidden by the foliage, but propagators now seem to be selecting from forms that display their flowers better. Hidden or not, the birds seem to enjoy them.

Origin NSW, particularly the south coast.
Size 2.5–3 m and spreading a little wider; sometimes much larger in native habitats.
Soil Adapts to the widest range of soils with no more than moderate drainage.
Climate Temperate to tropical; not severe frosts.
Aspect At least half sun.
Water Liberal to moderate; good resistance to dryness.
Feeding Accepts moderate applications of most complete fertiliser or animal manures.
Pruning May be pruned to thin forward growth and display flowers.
Propagation Seed or cuttings.
Possible Problems Usually trouble free.

Melaleuca incana

Melaleuca incana

Grey Honey Myrtle

FAMILY MYRTACEAE

This shrub is grown mainly for the graceful form of its pendulous outer branches and softness of grey foliage. The hairy grey lance-shaped leaves are to 1.5 cm long and the cream brushes of flower are to about 3 cm long and are carried on outer branchlets during winter to late spring. There is also a dwarf bushy compact blue-grey form available sold as 'Velvet Cushion'.

Origin Southern WA, usually in damp peaty sand.
Size 2.5–3.5 m high and a similar width. 'Velvet Cushion' grows to 1 m high and is a little wider.
Soil Although from moist areas, seems best in at least moderately drained sandy to medium loam.

Melaleuca laterita

Climate Cooler temperate to temperate.
Aspect At least half sun or lightly dappled sunlight.
Water Moderate to liberal with fair resistance to dry periods.
Feeding Slow release or organic fertilisers used sparingly in spring.
Pruning Needs light to moderate pruning after flowering to encourage the more attractive new growth. 'Velvet Cushion' needs branchlets carefully lifting to prune out dead under branches (use gloves against spiders).
Propagation Seed or cuttings; naturally the latter for 'Velvet Cushion'.
Possible Problems Leaf-webbing caterpillars; 'Velvet Cushion' may also die out, at least in patches, due to botrytis-type fungus during hot humid conditions.

Melaleuca laterita

(syn. *Melaleuca crassifolia*)

Robin Red Breast

FAMILY MYRTACEAE

A small shrub with blue-green linear leaves only about 1.5 cm long. Broad cylindrical orange-red flower spikes are carried back from the branch tips mainly during late spring to late summer. Tip-flowering forms are now being selected for propagation. Unlike *Melaleuca fulgens*, it does not demand keen drainage.

Origin WA, from Perth south to Albany but mainly in the more swampy areas.
Size 1.5–1.75 m and nearly as wide.
Soil Although from claypan moist areas it does well in composted sand with good drainage, therefore can be combined with the greater range of native plants.
Climate Cooler temperate to semi-tropical and generally tolerant of wet summer east coast conditions.
Aspect Full sun preferable in cooler temperate regions, lightly broken sunlight in warm climates.
Water Moderate to liberal; accepts normal dry spells.
Feeding Sparingly with complete organic or slow release fertilisers, or poultry manure.
Pruning May be pruned moderately after flowering.
Propagation Seed or cuttings.
Possible Problems Usually trouble free except perhaps for tip-webbing caterpillars which carry out their own form of tip pruning.

Melaleuca linariifolia

Snow in Summer

FAMILY MYRTACEAE

The species is a small bushy tree with very slender bright green linear to lance-shaped leaves to 3.5 cm long. It is spectacular in mid- to late summer or in Sydney/Brisbane in late spring when it is covered with what from a distance looks like a blanket of cottonwool. At close range the individual florets have creamy white feather-like arrangements of stamens. There is a dwarf form, 'Snowdrift', which reputedly grows no more than 2 m.

Origin Qld and NSW.
Size To 10 m high, about 5 m wide.
Soil Most reasonably moist garden soils.
Climate Cooler temperate to tropical. Does well in Melbourne Botanic Gardens.
Aspect Needs full sun to reach its full potential.
Water Liberal but accepts periods of dryness.
Feeding Sparingly with slow release or organic fertiliser.
Pruning Little required as a tree; dwarf

Melaleuca linariifolia

Melaleuca linariifolia (typical growth)

form may be pruned back as flowers finish.
Propagation Seed or particular forms from cuttings.
Possible Problems Usually trouble free.

Melaleuca megacephala

FAMILY MYRTACEAE

An erect strong growing shrub with glossy light greyish green ovate leaves about 2.5 cm long; the numerous slender stems carry terminal pompom-like clusters of pale yellow flowers in spring.

Origin WA, east of Darling Range and south towards Esperance.
Size 2–3 m high, about 1.5 m wide.
Soil Adapts to sandy or medium loam; likes fairly damp conditions but not soggy nor too well-drained.
Climate Temperate to semi-tropical; stands all but severe frost.
Aspect Best in open sunny situation.
Water Liberal to moderate.
Feeding Slow release or complete organic is safest, after spring flowering.
Pruning After flowering, cut well back but without denuding of foliage.
Propagation Seed or cuttings.
Possible Problems Usually trouble free.

Melaleuca megacephala

Melaleuca quinquenervia

Broad Leaf Paper Bark

FAMILY MYRTACEAE

An adaptable tree that will accept poorly drained soil. Useful where a tall screen is needed or for the beauty of its papery white bark and cream flower brushes. Leaves are lance shaped, to 9 cm long, or occasionally broader towards the top (oblanceolate), and unmistakable from other species by the parallel veins responsible for the species name (quinquenervia = five straight veins). The 8 cm long cylindrical brushes of flower come mainly in spring and autumn.

Origin Qld, NSW, New Guinea and New Caledonia.
Size Average height about 10–12 m and 4–5 m wide, but occasional specimens to 25 m occur.
Soil Very adaptable, from well-drained sandy loam to heavy wet clay.
Climate Tropical to temperate areas without severe frost.
Aspect At least half but ideally full sun.
Water Liberal welcome, but good resistance to dryness when established.
Feeding Moderate dressing of complete or slow release fertiliser or animal manure.
Pruning Only if necessary to reduce size.
Propagation Seed.
Possible Problems Usually trouble free.

Melaleuca quinquenervia

Melaleuca squarrosa

FAMILY MYRTACEAE

A handsome shrub displaying deep glossy green oval leaves less than 1 cm long arranged in two opposite rows and during late spring to early summer numerous erect cylindrical cream brushes 4-5 cm long with a sweet perfume.

Origin SA, Tas., Vic. and NSW, often found in moist peaty heathland.
Size 2.5-3 m high and as wide.
Soil Prefers moist sandy to medium loam with compost added but also adapts to average garden soil.
Climate Cooler temperate to semi-tropical.
Aspect Preferably lightly broken but also accepts full sunlight.
Water Liberal but tolerates dry periods.
Feeding Light dressings of organic or slow release fertiliser or animal manure in spring providing soil is moist.
Pruning More attractively compact if pruned back by at least one-third after flowering.
Propagation Seed and possibly from cuttings.
Possible Problems Usually trouble free.

Melaleuca steedmanii

FAMILY MYRTACEAE

A very showy melaleuca with resemblance to *Melaleuca fulgens*, with leaves fractionally shorter (about 2 cm), broader (narrow elliptical), and a little greyer. Also the bright scarlet brushes come earlier, from late winter into spring and birds are greatly attracted to them.

Melaleuca squarrosa

Origin WA, Irwin and Austin districts east of Geraldton in granite but often damp soil.
Size About 1 m high and a little wider.
Soil Preference for gravelly or coarse sandy loam with compost or peat added, but accepts all but toughest clay, well-drained or on damp side. Maintain a leafy surface mulch.
Climate Cooler temperate to semi-tropical but severe frost may damage early flowers.
Aspect At least half sun.
Water Moderate to liberal; tolerates dry periods.
Feeding Sparingly, preferably with complete organic or slow release fertiliser.
Pruning Prune back after flowers and for more compact growth tip prune new shoots.
Propagation Seed or cuttings.
Possible Problems Usually trouble free.

Melaleuca thymifolia

Thyme-leafed Honey Myrtle

FAMILY MYRTACEAE

A compact rounded small shrub, soft in appearance. It has small greyish green slender elliptical leaves about 1 cm long and mauve flowers with a silky incurving arrangement of stamens; these are carried mainly from spring to late summer.

Origin Qld and NSW, mainly in coastal areas.
Size 1-1.5 m and nearly as wide.
Soil Sandy to medium clayey loams with leaf mulch added. It accepts damp slow drying soils but not permanently soggy areas.

Melaleuca steedmanii

Melaleuca thymifolia

Climate Cooler temperate to semi-tropical; good frost tolerance.
Aspect Full sun to lightly broken sunlight preferable but a specimen in

the Melbourne Botanic Gardens continues to flower well with only a few hours late afternoon sun.
Water Liberal if possible but accepts dry periods.
Feeding Sparingly with slow release or organic fertiliser.
Pruning Prune back by at least one-third in late summer when flowers are finishing.
Propagation Seed or cuttings.
Possible Problems Reasonably trouble free.

Micromyrtus ciliata

FAMILY MYRTACEAE

A small shrub with tiny heath-like leaves and, from late winter into summer, minute cupped flowers that open white then turn a dull red as they age.

Origin SA, Vic. and NSW.
Size Several forms: the taller which is 0.75–1 m high and to 1.5 m wide, and a prostrate form of similar width but only to about 15 cm high.
Soil Most well-drained soils except heavy clay unless the latter is well broken with compost, gritty sand, etc.
Climate Cooler temperate to semi-tropical; frost tolerant.
Aspect Full to dappled sunlight.
Water Liberal to moderate with occasional dry periods.
Feeding Sparingly with slow release or organic fertiliser.
Pruning May be pruned lightly after flowering.
Propagation Cuttings.
Possible Problems Usually trouble free.

Mimulus repens

FAMILY SCROPULARIACEAE

An appealing little ground carpeter with soft green lance-shaped leaves 2–4 cm long and upfacing soft silvery lilac flowers about 3 cm across with a circular white zone surrounding the throat. These occur mainly during late summer and autumn.

Origin All States of Australia and New Zealand.
Size 8–10 cm high, spreading about 70 cm.
Soil Ideally a moist peaty sand, appreciates a little seepage, but adapts to most not too dry soils.

Micromyrtus ciliata

Climate Semi-tropical to cool; generally regarded as cool climate plant but the one illustrated was growing well at Ferndale near Nambour, Qld.
Aspect About half sun to lightly broken sunlight or, in cool areas, full sun.
Water Liberal.
Feeding Light dressing of complete plant food when planting out.
Pruning Not applicable.
Propagation Division
Possible Problems Seems trouble free where moisture is available.

Mirbelia oxylobioides

FAMILY PAPILIONACEAE

A very adapatable dense low growing shrub but with upright stems carrying closely set greyish green oblong leaves about 1 cm long and during late spring–early summer terminal clusters of orange pea flowers with a reddish blotch at the base of the standard; as the species name implies 'like oxylobium' (in this case *Oxylobium ellipticum*).

Origin Vic. and NSW.
Size 60–90 cm high, to about 1.5 m wide.
Soil Adapts to most well-drained soils.
Climate Cool to semi-tropical with good frost and seacoast tolerance.
Aspect Sun or light shade; the latter preferable in warmer districts.
Water Liberal to moderate; good resistance to dry spells.
Feeding Sparingly with slow release or other complete plant food.
Pruning May be pruned back to about half as flowers finish.
Propagation Cuttings or scarified seed.
Possible Problems Usually trouble free.

Mirbelia rubiifolia

FAMILY PAPILIONACEAE

A small open shrub with very sparse

Mimulus repens

whorls of slender lance-shaped leaves 2–3 cm long and heads of rosy purple pea flowers 1–1.5 cm across during spring. Interesting trailer or rock garden plant.

Origin NSW, mainly central to southern coast, in sandstone coastal heathland areas.
Size About 50 cm high, to 75 cm wide.
Soil Well-drained composted sand or light loam with leafy mulch or pebble scree.
Climate Temperate.
Aspect Lightly broken sunlight to dappled shade.
Water Liberal to moderate.
Feeding Sparingly if at all with blood and bone or other organic plant food.
Pruning Prune back lightly after flowering.
Propagation Sacrified seed.
Possible Problems Usually short lived and resents high nitrogen fertilisers, especially sulphate of ammonia.

Mirbelia oxylobioides

Mirbelia rubiifolia

Myoporum debile

'Amulla'

FAMILY MYOPORACEAE

A prostrate shrub for groundcover. The dense 4–7 cm long slender lance-shaped foliage is a deep bluish green. Small pink to white flowers along the trailing stems during spring and summer are followed by 7–8 cm long white berries heavily flushed rosy red, with some resemblance to tiny apples.

Origin Qld and NSW.
Size 15–30 cm high and spreading to about 1 m.
Soil Preferance for well-drained but reasonably moist light loam but adapts to most garden soils. Retard weed growth with leafmould or woodchip mulch while plants are establishing.
Climate Cooler temperate to semi-tropical; moderate frost resistance.
Aspect At least half sun.
Water Liberal to moderate but reasonable drought resistance when established.
Feeding Slow release or organic fertiliser, mainly in spring.
Pruning Tip prune if needed to encourage further branching or to check ascending growth.
Propagation Cuttings.
Possible Problems Scale, indicated by sooty mould on foliage or stems.

Myoporum debile 'Amulla'

Myoporum floribundum

Myoporum parvifolium

Myoporum floribundum

Boobiala

FAMILY MYOPORACEAE

A large shrub with gracefully pendulous slender linear leaves, 8–10 cm long hanging like green rain from almost horizontal branches; the branches during spring are tightly clustered with tiny five-petalled starry white flowers.

Origin Vic. and NSW.
Size To 3 m high and as wide.
Soil Moist well-drained soils.
Climate Cooler temperate to semi-tropical.
Aspect Full to lightly broken sunlight.
Water Liberal to moderate; once established has good resistance to long dry periods.
Feeding Slow release or complete organic plant food during mid- to late spring.
Pruning May be pruned lightly after flowering if necessary.
Propagation Seed or cuttings.
Possible Problems Occasionally found with scale but usually trouble free.

Myoporum parvifolium

(syn. *Myoporum hymile*)

Creeping Boobiala

FAMILY MYOPORACEAE

A contour-following trailing plant making a carpet of bright green slightly warted and serrated club-shaped leaves 2–3 cm long; during late spring and summer the slender branches are closely studded with upfacing starry white flowers barely 1 cm across with pale lilac spots and pink flushed buds. There is also a form with slightly stubbier foliage that turns purple-bronze during winter. Both make attractive and effective self-layering groundcovers or trailers.

Origin Vic., Tas., SA and WA.
Size Mats to about 20 cm high and spreads 1 m or more. There are upright forms less suitable as groundcovers.
Soil Most well-drained soils, preferably started with woodchip or leafy mulch.
Climate Cool to semi-tropical; good frost resistance.
Aspect At least half sun.
Water Liberal to moderate; when established has good drought resistance.
Feeding Sparingly with slow release or organic fertilisers.
Pruning If necessary tip prune to thicken and cut back unwanted growth.
Propagation Seed, cuttings or layers.
Possible Problems Check occasionally for scale which, if unchecked, may defoliate.

Nelumbo nucifera

Nelumbo nucifera

Sacred Lotus

FAMILY NELUMBONACEAE

A beautiful herbaceous water plant often featured in Asian art. It has impressive but graceful long stemmed and slightly fluted soft green disc-shaped leaves, 30-60 cm across, and elegant slender oval buds clapsed in long lime green bracts. The latter unfold to frame an exquisite arrangement of soft pink petals surrounding a large flat topped golden centre housing the slightly protruding nut-like embryo seeds. These flowers can measure up to 20 cm across. After maturity the centre becomes a woody cone honeycombed with from ten to forty empty seed cavities which are prized for dried flower arrangements. The 2 cm long pepper-pot-shaped seeds and the rhizomes were eaten by Northern Australian Aborigines and North American Indians; they still are eaten in some Asian countries.

Origin North Australia, New Guinea, Southern Asia, Central America and Florida.
Size 1-1.3 m above water, about 1.5 m wide then gradually spreading.
Soil Good rich composted soil in a submerged container or at the base of a pond, water preferably 30-40 cm deep and slow moving.
Climate Temperate to tropical.
Aspect Sunny.
Feeding Slow release fertiliser may be added to soil but too much increase algae in ponds.
Pruning Spent leaves may be removed.
Propagation Seed or by division.
Possible Problems Usually trouble free.

Nematolepis phebalioides

FAMILY RUTACEAE

An interesting and attractive small woody and fairly open shrub suitable for container growing. It has clean cut, deep green and fairly rigid oval leaves to 1.5 cm long, margins slightly rolled downwards, and 3 cm long pendulous bright glossy red tubular flowers with a greenish yellow tip. These come from the leaf axils during spring.

Nematolepis phebalioides

Origin WA, Stirling Ranges and east to Esperance.
Size About 1 m high, to 1.5 m wide.
Soil Well-drained, light, open, preferably with a gravely scree.
Climate Temperate; frost tolerant.
Aspect Full to lightly broken sunlight.
Water Moderate to liberal; good drought tolerance when established.
Feeding Sparingly with slow release or complete organic fertiliser.
Pruning Pruninge lightly after flowering.
Propagation Cuttings.
Possible Problems Resents hot wet summer conditions or root disturbance.

Nuytsia floribunda

WA Christmas Bush

FAMILY LORANTHACEAE

A spectacular tree, related to mistletoe, which appears to at least start its life as a parasite on roots of other plants. Leaves are stiff, slender, to about 10 cm long, usually with margins rolling downwards, but variable. The showy flowers come in clusters of up to twelve tightly packed spikes radiating from the tip of each branchlet; they are bright glossy green in bud and open to a deep golden yellow, usually about Christmas time.

Origin WA, mainly south-western corner.
Size Usually 4-5 m high and 3-4 m wide but can grow to twice this size.
Soil Fairly well-drained gritty loam or well-composted sand.
Climate Cooler temperate to semi-tropical; good frost tolerance.
Aspect Open sunny position in presence of other lower plants.
Water Liberal to moderate; withstands dry periods.
Feeding Not fully established but a complete plant food not too rich in nitrogen would be necessary if establishing couch grass as a host.
Pruning If necessary for shaping; results of heavy pruning at time of writing are uncertain.
Propagation Seed or root cuttings; it has been established by sowing couch with nuytsia seed then keeping the couch reasonably trimmed back. Grafting may be a possibility.
Possible Problems Seems to be mainly the dependence on a suitable host.

Nymphea gigantea

Tropical Water Lily

FAMILY NYMPHACEAE

SEE PAGE 173

Like the lotus, this is another plant beautifully presented by nature but in this case the smaller and glossier green and slightly toothed leaves, 20-50 cm in diameter, are provided with more

Nuytsia floribunda

flexible stems so that they can float on the water; unlike the cooler climate counterparts the flowers stand erect above the surface, are usually blue, sometimes white, and they are 15–20 cm or occasionally 30 cm across.

Origin NT, Qld and warmer coastal NSW.
Size The creeping rhizomes eventually make clumps 2–3 m across but they can be confined for several years to a tub or pool 50–60 cm across.
Soil Rich well-composted loam at base of pool container, soaked thoroughly before submerging to minimise floating out of organic or other matter.
Climate Warm; where water does not become too chilly.
Aspect At least half but preferably full sun.
Water Needs to be at least 60 cm deep.
Feeding Slow release may be added to soil mix but too much nitrogen (in conjunction with light) can cause algae problems.
Pruning Remove old leaves for better appearance.
Propagation Division of rhizomes during winter.
Possible Problems Small beetles sometimes attack foliage, usually controlled by pushing leaves under water, making the pests accessible to fish which are needed to also control mosquitoes.

Olearia dentata

Daisy Bush

FAMILY ASTERACEAE

A small shrub with toothed, stiff leathery leaves, 3–5 cm long. During spring or early summer it carries white or mauve daisies with comparatively large centres and long stems.

Origin Vic. and NSW.
Size To 1.5 cm high and as wide.
Soil Adapts to most well-drained garden soils.
Climate Cool to warmer temperate.
Aspect Half to lightly broken sunlight.
Water Liberal to moderate but good drought resistance.
Feeding Slow release or organic fertiliser applied moderately during early summer is acceptable.
Pruning Prune back after flowering and remove old flower heads to keep the plant attractive.
Propagation Cuttings or seed.
Possible Problems Fairly trouble free.

Olearia floribunda

Daisy Bush

FAMILY ASTERACEAE

A small shrub with closely set oblong deep green leaves only about 3 mm long and massed with fine petalled white daisies 1–2 cm across during spring and early summer. There is also a mauve to lavender form.

Origin SA, Tas., Vic. and NSW.
Size 1–1.7 m high and as wide.
Soil Adaptable; found growing naturally in a variety of soil types with moderate to keen drainage.
Climate Cool to warmer temperate; good frost resistance.
Aspect Preference for lightly broken sunlight to dappled shade.
Water Liberal to moderate; good drought resistance.
Feeding Light sprinkling of complete or slow release fertiliser as new growth begins is acceptable.
Pruning Need pruning back after flowering to keep growth attractive.
Propagation Cuttings.
Possible Problems With overhead shelter or during very dry periods red spider mites may cause dull mottling of foliage.

Olearia floribunda

Olearia phlogopappa

Large Flowered Daisy Bush

FAMILY ASTERACEAE

Small shrub, more erect than most olerias with deep greyish green oblong to elliptical rough leaves to about 4 cm long. Showy comparatively broad petalled daisies 2–2.5 cm across mass the plant during spring. There are white, mauve, pink and lavender blue colour forms.

Origin Tas., Vic. and NSW.
Size 1.5–1.7 m high, about 0.75 m wide.
Soil All but the heaviest well-drained garden soils are suitable.
Climate Cool to semi-tropical; good frost resistance.
Aspect At least half sun.
Water Moderate; good drought resistance when established.
Feeding Slow release or organic fertiliser is acceptable when new growth starts in early summer.
Pruning Prune back after flowering to keep plants attractive longer; best renewed after a few years.
Propagation Cuttings from selected colour forms.
Possible Problems Leaf miner may cause squiggle like tracery on foliage but usually trouble free.

Olearia dentata

Olearia phlogopappa

Nymphea gigantea

Oreocallis wickhamii

Pandorea jasminoides

Orchids

SEE PAGE 221

Oreocallis wickhamii

(syn. *Embothrium wickhamii*)

Tree Waratah

FAMILY PROTEACEAE

A spectacular tree, initially pyramid shaped then after about twenty-five years developing a rounded crown. The dark glossy green slender oblong to lance shaped or occasionally slightly lobed leaves are 10–12 cm long then in spring it carries clusters of brilliant red flowers that in form could be described as between a large spider-flowered grevillea and a loosely packed waratah (it is related to both).

Origin North Qld rainforests.
Size Average size in fifteen years 7–8 m high, 3–4 m wide at head height; eventually to 15 m high or more.
Soil Prefers well-drained and composted light to medium loam.
Climate Temperate to tropical, tolerance to no more than moderate frosts; best on humid coast than drier inland areas.
Aspect Accepts initial shade, especially in tropical areas; elsewhere at least two-thirds sun with shelter from cold winds is preferable.
Water Liberal to moderate; established trees accept long periods of dryness.
Feeding Complete water soluble or slow release fertilisers speed early growth.
Pruning Only if necessary for shape.
Propagation Seed.
Possible Problems Resents soils overwet during the cold months.

Pandorea jasminoides

(syn. *Bignonia jasminoides*)

Bower of Beauty

FAMILY BIGNONIACEAE

A beautiful twining plant with large bright green glossy foliage divided into elliptical to broad lance-shaped leaflets 4–6 cm long. Clusters of large waxy soft pink flowers about 4 cm across with a deep maroon centre appear spasmodically, mainly through the warmer months.

Origin Northern NSW and Qld to South-East Asia.
Size Slowly climbs 2–3 m; sometimes more rapid and spreading in warm moist areas.
Soil Most well-drained garden soils, preferably well composted.
Climate Temperate and tropical, stands moderate frost.
Aspect Morning sun or lightly broken sunlight is preferable but accepts full sun where humidity is at least moderate.
Water Liberal to moderate.
Feeding Slow release fertiliser applied in spring will speed growth; apply a light scattering of superphosphate or basic complete fertiliser if overleafy and without flowers.
Pruning To control size or create extra bushiness.
Propagation Cuttings or by layering.
Possible Problems Usually trouble free.

Pandorea pandorana

(syn. *Tecoma australis*)

Wonga Wonga Vine

FAMILY BIGNONIACEAE

A lush and attractive climber with good dense foliage throughout the year. The large leaves are divided into ovate to oblong leaflets about 5 cm long, dark glossy green. During spring it is showy with cascades of creamy buff tubular flowers tinged purple at the tips in loose terminal panicles. There is also a popular and free flowering creamy white form, 'Snowbells', and a golden buff to light tan, 'Golden Showers'.

Origin Tas., Vic., NSW and Qld, both over rainforest trees with its nutrients and moisture coming via 'monkey

Pandorea pandorana 'Snowb

Pandorea pandorana

ropes', and as a low cover in open areas.
Size Outside the rainforest it can grow about 5 m in either direction but easily controlled.
Soil Any reasonably well-drained garden soil with compost added.
Climate Cooler temperate to tropical; good frost tolerance.
Aspect Grows well in light shade or full sun.
Water Liberal to moderate.
Feeding Slow release or complete fertiliser speeds growth, the latter is more supportive of flowering.
Pruning When necessary to control size. Downward arching of wandering canes subdues and encourages branching.
Propagation Cuttings or by inducing layering.
Possible Problems Usually trouble free.

Patersonia glabrata

Native Iris

FAMILY IRISACEAE

Unlike the other patersonias the narrow 15–20 cm leaves of this species clasp the base of the flower stem making a fan-shaped formation. They are also smooth except at the base. Flower stems are usually shorter than leaves, carrying three comparatively smooth silky petals usually purple but can be soft lavender to pale violet, appearing during spring to early summer, usually lasting less than one day.

Origin Eastern Australia, mainly coastal.
Size As above; clumps usually 20–40 cm wide.
Soil Fairly well-drained composted soil, light loam or heavier loam lightened with leafmould.
Climate Cool to semi-tropical; frost tolerant.
Aspect Full to broken sunlight.
Water Liberal to moderate.
Feeding Rarely necessary but they accept complete organic or slow release fertiliser in spring.
Pruning Clean off old foliage and flower stems when flowering ceases.
Propagation Seed which usually germinates well.
Possible Problems Rarely affected.

Patersonia occidentalis

(syn. *Patersonia longiscapa*)

Similar to *Patersonia glabrata*, but usually with longer foliage, and more freely produced large deep to pale violet flowers.

Patersonia sericea

Patersonia glabrata

Patersonia occidentalis

Origin WA.
Size Height 30–40 cm, clumps to 40 cm wide.

Cultivation as for *P. glabrata* but prefers a moist soil with reasonable drainage.

Patersonia sericea

Native Iris

FAMILY IRIDACEAE

Clumps of slender but firm textured reed-like leaves, 30–40 cm long are covered with a thin film of minute hairs giving a slightly silky appearance. The flower stems, which carry rather flimsy flowers with three broad oval crepe-textured violet petals 2–3 cm long, are also silky. It is the largest of the patersonias, flowering during spring and summer. Flowers last less than a day but are freely produced.

Origin Vic., NSW and Qld.
Size As above; clumps rarely exceed 30–40 cm across.

Requires as for *Patersonia glabrate*.

Persoonia pinifolia

Pine-Leafed Geebung

FAMILY PROTEACEAE

A graceful and decorative shrub, its branches softly misted by dense but fine pine-like leaves about 5 cm long. During summer the outer branchlets terminate in spikes of small yellow flowers, which become pendulous under the weight of succulent oval fruits like small green grapes flushed bronze.

Persoonia pinifolia

Origin NSW, particularly in Hawkesbury sandstone areas.
Size 3 or occasionally 4 m high and nearly as wide.
Soil Standstone soils with plenty of leafmould or lighter well-drained clay-loams.
Climate Temperate or semi-tropical; frost tolerant.
Aspect Full of dappled sunlight.
Water Liberal to moderate; good drought resistance when established.
Fertiliser Sparingly if at all, using slow release or organic types.
Pruning Improved by light pruning as berries finish.
Propagation Seed or cuttings but unfortunately both are difficult.
Possible Problems Usually trouble free.

Persoonia levis

Broad Leafed Geebung

FAMILY PROTEACEAE

An erect open shrub, trunk and limbs

Persoonia levis

Petrophile canescens

covered with flakes of silky reddish black bark, and broad teardrop-shaped smooth bright green leaves 10–18 cm long and 6–8 cm wide, broadly rounded at the apex. The yellow flowers appear sparse as they open only a few at a time during early to mid-summer and are followed by bright green fruits to 2 cm long.

Origin Vic. and NSW.
Size 3–5 m high and about 2 m wide.
Soil Composted sand to medium loam, well-drained.
Climate Cooler temperate to semi-tropical; good frost tolerance.
Aspect Full to dappled sunlight.
Water Liberal to moderate but with fair drought resistance.
Feeding Sparingly with organic or slow release if any. Leafmould usually provides sufficient nutrient.
Pruning Lightly only if needed; rarely recovers well from heavy pruning.
Propagation Seed or cuttings (with difficulty).
Possible Problems Except for occasional fungus smut it is trouble free.

Petrophile canescens

Conesticks
FAMILY PROTEACEAE

An upright slender shrub with leaves to about 10 cm long, finely divided into spreading soft needle-like segments, young growth often golden bronze maturing dark green. During late spring and early summer, globular creamy gold flower clusters are displayed at the tips of branches. These are followed by metallic grey woody seed cones 3–4 cm long.

Origin Qld and NSW.
Size To 3 m high; much shorter in exposed coastal areas.
Soil Grows on a variety of well-drained soils with preference for gravelly sand with a leafy mulch.
Climate Cooler temperate to semi-tropical; frost tolerant.
Aspect Full to broken sunlight.
Water Moderate to liberal; withstands dry periods.
Feeding Sparingly with slow release or organic fertiliser.
Pruning May be rejuvenated and have base branching encouraged by pruning back to about half once every few years before new growth begins.
Propagation Seed which may be slow and germinate spasmodically.
Possible Problems Usually trouble free.

Phebalium

Although flowers are smaller and in larger clusters, botanically this genus has the same arrangement of five petals and ten free stamens as *Eriostemon* except that the petals of the latter are folded in bud (imbricate) and those of *Phebalium* are valvate (edges meet but do not overlap).

Phebalium squamulosum

Phebalium squamulosum

FAMILY RUTACEAE

An erect reasonably dense shrub with deep green linear to elliptical leaves from 1-2.7 cm long and dense clusters of small yellow flowers with prominent stamens mostly in spring.

Origin Eastern Australia, especially along sandstone regions.
Size Variable from 1-3 m in sheltered areas and 1-2 m wide.
Soil Needs well-drained, preferably elevated and mulched sandy to light loam.
Climate Cooler temperate to semi-tropical; moderate frost tolerance.
Aspect Full to dappled sunlight.
Water Liberal to moderate; good tolerance to dry periods.
Feeding Sparingly with organic or slowly soluble fertiliser in spring.
Pruning Moderately after flowering.
Propagation Cuttings.
Possible Problems Usually trouble free except that they can die out suddenly, probably with root fungi.

Phebalium squamulosum sub sp. *parvifolia*

Phebalium squamulosum sub. sp. *parvifolia*

(syn. *Phebalium parvifolia*)
FAMILY RUTACEAE

Similar to *Phebalium squamulosum* but with tiny foliage and it is usually more compact.

Origin Eastern Australia.
Size Height to about 1 m and similar width.
Soil Well-drained and composted sandy to light loam.
Climate Cooler temperate to semi-tropical; frost tolerant.
Aspect Full to lightly broken sunlight.
Water Liberal to moderate; good drought tolerance when established.
Feeding Very sparingly with organic or slow release fertiliser in spring.
Pruning Prune back lightly after flowering.
Propagation Cuttings.
Possible Problems Sometimes dies suddenly, probably with root fungus but worth trying.

Phebalium whitei

FAMILY RUTACEAE

A small slender shrub carrying dark green slender oblong foliage, to 6 cm long, and, in spring, clusters of clear bright yellow flowers which are unusually large for *Phebalium*. Ironically it is the most difficult of the genus to maintain.

Origin Qld.
Size About 50 cm high and as wide.
Soil Well-drained composted gritty loam seems preferable.
Climate Temperate to semi-tropical; frost resistant.
Aspect Broken sunlight or morning sun.
Water Liberal to moderate.
Feeding Very sparingly with organic plant food.
Pruning Lightly after flowering.
Propagation Cuttings.
Possible Problems Usually short lived, subject to root fungus.

Phebalium whitei

Philotheca salsolifolia

(syn. *Philotheca Australis*)
FAMILY RUTACEAE

Like its close relatives, *Phebalium* and *Eriostemon*, this genus also has five petals and ten stamens but the latter are joined rather than free to their base. It is a small twiggy shrub with slender heath-like leaves, to 1 cm long, and starry lavender blue flowers, usually occurring singly at the ends of the branchlets during spring.

Origin Qld and NSW, especially in sandy coastal areas.
Size Height to 1.5 m usually less than 1 m, to 1 m wide.
Soil Composted sand, preferably reasonably moist but well drained.
Climate Cooler temperate to semi-tropical; good frost tolerance.
Aspect Lightly broken sunlight or full sun with base protected by other small plants or a leafy mulch.
Water Liberal to moderate; in cultivation it can die out if dry for long periods.
Feeding Very sparingly with organic or slow release fertiliser in spring.
Pruning Prune back lightly as flowers finish.
Propagation Cuttings but it's not the easiest to strike.
Possible Problems Except for resentment of prolonged dryness and a possible leaf-tip-rolling caterpillar it is usually trouble free.

Philotheca salsolifolia

Pileanthus peduncularis

Pimelea glauca

Pileanthus peduncularis

Copper Cups

FAMILY MYRTACEAE

A showy and free flowering shrub with fine heath-like foliage and from late winter to late spring, generously clustered with five-petalled saucer-shaped flowers to about 5 cm across, the silky petals orange-red to copper colour with bronze-purple centre and stamens. When flowers finish they fall to leave a buff daisy-like calyx.

Origin WA, on the sand plain between Geraldton and Perth and 200–300 km west.
Size 0.75–1.5 m and as wide.
Soil Well-drained composted sand or gritty loam.
Climate Temperate to semi-tropical; frost tolerant.
Aspect Preferably full sun for at least two-thirds of the day.
Water Moderate.
Feeding Moderately with slow release or complete organic fertiliser in late autumn and lightly again in spring; only if reasonable water is available.
Pruning May be pruned lightly after flowering.
Propagation Seed and cuttings.
Possible Problems A good reliable shrub in Perth gardens. It appears that it would be adaptable to eastern Australian conditions but there is so far no information on its acceptance of wet summers.

Pimelea ferruginea

FAMILY THYMELAEACEAE

A compact and attractive garden shrub, with or without flower. The small oval leaves are deep glossy green, about 1 cm long, closely and neatly arranged in two pairs of opposite rows. The profuse dome-shaped head of flowers are 3–4 cm across, originally soft rose pink but nurserymen seem to be selecting a brighter and harder pink strain. They flower throughout spring.

Origin WA, mainly lower south-west in limey coastal to slightly acid granite soils.
Size About 1 m high and at least as wide.
Soil Adapts to most well-drained garden soils.

Climate Cooler temperate to semi-tropical; frost tolerant.
Aspect At least half sun.
Water Liberal to moderate.
Feeding Light dressings of complete fertiliser, animal manure, other organic or slow release fertiliser.
Pruning Prune back by about one-third after flower.
Propagation Cuttings.
Possible Problems Seems trouble free in most areas.

Pimelea glauca

FAMILY THYMELAEACEAE

A small compact and spreading shrub with fine, rather lax, linear to lance-shaped leaves, to 1 cm long, and heads of small creamy white flowers in spring and summer.

Origin SA and all eastern States including Tas.
Size 30–60 cm high, to about 75 cm wide.
Soil Most reasonably well-drained garden soils.
Climate Cooler temperate to semi-tropical; frost tolerant.
Aspect Full to lightly broken sunlight.
Water Moderate to liberal.
Feeding Slow release or organic fertiliser when new growth commences.
Pruning May be pruned back lightly or to half as flowers finish.
Propagation Cuttings.
Possible Problems Rarely affected.

Pimelea ferruginea

Pimelea linifolia

Rice Flower
FAMILY THYMELAEACEAE

A small erect shrub with two pairs of opposite rows of usually small linear leaves to about 3 cm long but can be broader, almost ovate. Terminal heads of white flowers, sometimes with pink flushed buds occur mainly during winter and spring but also occasionally during late summer and autumn. Foliage is suspected of being poisonous to stock.

Origin SA and eastern States including Tas.
Size 40–60 cm high and about 40 cm wide.
Soil Occurs frequently in sandstone areas but adapts to most well drained soils.
Climate Cooler temperate to tropical; frost tolerant.
Aspect Light shade or full sun.
Water Liberal to moderate; when established withstands dry periods.
Feeding Sparingly only with organic or slow release fertiliser.
Pruning Tip prune when young to encourage bushiness. May also be pruned back to about half after flowering.
Propagation Cuttings or seed.
Possible Problems Usually trouble free.

Pimelea ligustrina

Large Leafed Rice Flower
FAMILY THYMELAEACEAE

An erect open shrub with smooth elliptical to lance-shaped mid-green leaves to 5 cm long and long stemmed pompom-like clusters of white flowers during summer. Useful for shaded areas.

Origin SA and all eastern States.
Size To about 2 m high and 1.5 m across.
Soil Adapts to a wide range of soils, preferably with a leafy mulch.
Climate Cooler temperate to semi-tropical.
Aspect Dappled to all but very dense shade.
Water Liberal to moderate.

Pimelea linifolia

Feeding Organic, slow release fertiliser or fowl manure in moderation during spring.
Pruning Prune back lightly after flowering or more heavily in winter.
Propagation Cuttings:
Possible Problems Usually trouble free.

Pimelea ligustrina

Pimelea nivea

Pimelea spectabilis

Pimelea nivea

FAMILY THYMELAEACEAE

A small spreading plant but with numerous upright downy white stems, with widely spaced opposite small rounded leaves to 1.5 cm long, dark green above, grey beneath and terminating in heads of white flowers during summer.

Origin Tas.
Size 1–1.5 m high and nearly as wide.
Soil Most fairly well-drained soils, preferably with compost added or a leafy mulch.
Climate Cool and temperate; good frost tolerance.
Aspect Best in light shade.
Water Liberal to moderate; resents prolonged dryness.
Feeding Sparingly with slow release or, organic fertiliser or fowl manure in spring.
Pruning Stems may be cut back to about two-thirds after flowers finish.
Propagation Tip cuttings.
Possible Problems Die back in hot dry areas.

Pimelea rosea

Pink Rice Flower

FAMILY THYMELAEACEAE

A small slender and fairly open shrub with lance-shaped leaves to 1.5 cm long and terminal heads of flower to about 30 cm across, usually with the petals a bright cerise pink, occasionally pale pink or white. Flowers late winter to late spring.

Origin Southern WA.
Size About 0.75 m high and nearly as wide.
Soil Composted sand to medium loam, in all cases a leafy mulch is preferable and good drainage.
Climate Cooler temperate to semi-tropical; frost tolerant.
Aspect Lightly broken sunlight or some shade during the hottest part of the day is preferable.
Water Liberal to moderate; reasonable tolerance to dryness when established.
Feeding Light dressings of complete, organic, or slow release fertiliser or blood and bone in spring as flowers are fading.
Pruning Flowers carrying stems may be shortened back to about one-third after flowering.
Propagation Cuttings.
Possible Problems Usually trouble free.

Pimelea spectabilis

Bunjong

FAMILY THYMELAECEAE

A small fan-shaped shrub like a sheaf of erect slender branches with closely set thin but rigid slender lance-shaped to oblong leaves to 3 cm long. During spring these terminate in flower clusters made spectacular by regular placement of the long slender tubular florets. These radiate like creamy white spokes of a wheel from a hub of unopened crimson buds, then open their slender petals to make a lace ice-pink circular frame. This floral arrangement can also be yellow, deep pink or white and is 4–5 cm across and kept in place by four large supporting bracts.

Origin Southern WA.
Size 0.75–1.5 m high and almost as wide.
Soil Composted sand or light to medium loam with good drainage.
Climate Cooler temperate to semi-tropical.
Aspect Best in lightly broken sunlight or morning sun.
Water Liberal to moderate; reasonable tolerance to dry periods.
Feeding Moderate application of standard complete, slow release, fowl manure or other organic fertilisers in late spring.
Pruning Flower carrying stems can be cut back to about one-third as flowers are finishing.
Propagation Cuttings
Possible Problems Usually trouble free.

Pimelea rosea

Pittosporum phillyroides

Bitter Bush,
Weeping Pittosporum

FAMILY PITTOSPORACEAE

A small tree with pendulous outer branchlets and willowy linear to lance-shaped leaves 8–11 cm long. The creamy yellow spring to summer flowers are followed by sprays of egg-shaped orange berries along branchlets behind new growth. A good light shade or street tree for dry inland areas.

Origin Inland areas of all mainland States.
Size 6–8 m and 4–5 m wide above head height.
Soil Most well-drained soils.
Climate Cool temperate to tropical. Frost tolerant.
Aspect Open, sunny.
Water Moderate only; once established is drought tolerant.
Feeding Complete fertiliser, fowl manure or slow release applied sparingly just beyond dripline of tree when soil is moist.
Pruning For shape or in spring to rejuvenate at expense of berries.
Propagation Seed.
Possible Problems Occasional scale but usually trouble free.

Pittosporum phyllyroides

Pittosporum rhombifolium

FAMILY PITTOSPORACEAE

An initially erect then later spreading small tree with bright glossy green leaves about 10 cm long. Its main feature is the mass of bright orange berries that follow the clusters of small cream summer flowers. These berries are globular, about the size of a pea (about 0.75 mm in diameter), glossy and

Pittosporum rhombifolium

Pittosporum undulatum

Platylobium formosum

Plectranthus graveolens

remain from late summer into late winter.

Origin Qld and NSW rainforests.
Sizee 8–10 cm high, at least twice as high in rainforests; for first ten years to about 2.5 m wide, then gradually to about 5 m.
Soil Most well-drained garden soils.
Climate Temperate to tropical; reasonable frost tolerance but happier in the more humid coastal regions.
Aspect At least half sun.
Water Liberal but once established withstands drought.
Feeding Water soluble plant foods in spring and summer speed initial growth; otherwise slow release, fowl manure or complete fertiliser sparingly.
Pruning Only if needed for shape.
Propagation Seed.
Possible Problems Scale, indicated by presence of black sooty mould on stems and foliage.

Pittosporum undulatum

Native Daphne

FAMILY PITTOSPORACEAE

A quick growing umbrella-shaped tree with bright green elliptical leaves 8–12 cm long, terminal clusters of small fragrant creamy white bell-shaped flowers in spring, followed by dull orange-tan berries about 1 cm across. These fall and split exposing sticky pitch-covered seeds which can be a nuisance on paved paths or living areas. Inclined to become invasive in some coastal gullies but a handsome tree.

Origin Eastern States and SA.
Size 10 m high, spreading 7–8 m, occasionally larger.
Soil Any moderately drained garden soil.
Climate Cooler temperate to tropical; frost tolerant.
Aspect Anywhere except indoors.
Water Liberal but tolerance to long dry spells.
Feeding Not necessary except that feeding may curb its rapid root spread, then give it anything.
Pruning If needed to restrict growth; anytime.
Propagation Seed.
Possible Problems Leaves invariably carry small circular thickened indented areas caused by a minute leaf miner fly larvae. This is of little detriment and does not warrant chemical control, which in any case is almost impossible.

Pomaderris multi

Platylobium formosum

Flat Pea

FAMILY PAPILIONACEAE

A lax but fairly erect small shrub with stiff heart-shaped leaves about 5 cm long, dark green above, greyish beneath, and small clusters of yellow and red pea flowers during spring. Useful for shaded areas under trees.

Origin Eastern States of Australia, particularly east of Dividing Range.
Size 1–1.5 m high and as wide. There are more compact forms.
Soil Most well-drained soils but prefers a leafy surface mulch.
Climate Cooler temperate to semi-tropical.
Aspect Accepts full sun but useful under trees where too shaded for many other small shrubs.
Water Liberal to moderate; good tolerance to dry periods.
Feeding Sparingly only with organic complete plant food or blood and bone (sensitive to high nitrogen).
Pruning Improved by pruning back by about one-third after flower.
Propagation Scarified seed.
Possible Problems Leaf-miner caterpillar may leave brown blotched foliage.

Plectranthus graveolens

Musk

FAMILY LAMIACEAE

A soft wooded small shrubby perennial with thick downy felt-like, and aromatic, heart-shaped 3–5 cm long on square stems typical of the family. During summer slender spikes of small salvia-like lavender blue flowers appear with a slightly pouched lip.

Origin Eastern Australia and SA, mainly from coast to mountains, on rocky ledges where a little soil and leafmould has collected.
Size About 40 cm high and as wide but gradually spreading.
Soil Any well-drained soil or in shallow containers.
Climate Cooler temperate to semi-tropical, stands light frost only.
Aspect Any but dense shade.
Water Moderate; withstands dryness well.
Feeding Light application of complete fertiliser in spring.
Pruning Prune off old flower heads and in winter reduce any leggy growth. May be tip pruned later.
Propagation Seed but very easy from cutting, in shade but without plastic cover or high humidity.
Possible Problems Lower foliage eventually dies, better to start new plants from tip cuttings.

Pomaderris discolor

Pomaderris discolor

FAMILY RHAMNACEAE

Medium size shrub with elliptical to broad lance-shaped leaves 4–8 cm long, dark green above, downy and grey underneath and terminal clusters of small cream flowers in spring, often with petals absent.

Origin Eastern Australia, mainly in forests of temperate coastal areas.
Size 1–4 m high and about 2.5 m wide.
Soil Most well-drained loamy soils or sand with leafy mulch.
Climate Temperate to semi-tropical; some frost tolerance.
Aspect Dappled sunlight.
Water Liberal to moderate.
Feeding Sparingly with slow release or organic fertiliser in spring but satisfied with leafmould mulch.
Pruning Pruning as flowers finish encourages more compact growth.
Propagation Cuttings.
Possible Problems Occasional bud or tip growth attacked by webbing caterpillars.

Pomaderris multiflora

FAMILY RHAMNACEAE

A usually rounded and fairly dense shrub with elliptical leaves 5–10 cm long, dark green above, downy grey underneath with prominent brown veins and during spring very showy pale yellow flowers in terminal panicles that almost obscure the foliage.

Origin Eastern Australia, usually on well-timbered Dividing Range foothills and coast from eastern Vic. to Qld.
Size 1.5–4 m and 1.5–3 m wide.
Soil Well-composted sandy to medium loam with good drainage.
Climate Cooler temperate to semi-tropical; frost tolerance.
Aspect Lightly broken sunlight to dappled shade.
Water Liberal but accepts dry periods when established.
Feeding As for *Pomaderris discolor*.
Pruning May be pruned back as flowers finish.
Propagation Cuttings.
Possible Problems As for *P. discolor*.

Prostanthera

Belongs to family Lamiaceae (or previously more frequently listed as Labiacea meaning 'lipped') which also includes salvia, coleus, thyme, sage, mint, westringea, etc. All have lipped flowers, leaves opposite or in whorls, often with musky or spicy aroma. Most have squared rather than rounded stems, in some cases noticeable only on new growth. *Prostanthera* and *Westringeas* have very similar flowers, the distinguishing factor being that in *Prostanthera* the calyx is made up of two broad lobes like a smooth edged cowl or lamp holder at the back of the flower, whereas the *Westringea* calyx has five pointed lobes.

Prostanthera lasianthos

Prostanthera incisa

Prostanthera incisa

FAMILY LAMIACEAE

Small compact shrub with scalloped or toothed oval leaves to 1.5 cm long that are bright but matt green and deeply veined. Profuse clusters of bright lavender to pale violet flowers are carried during spring.

Origin Eastern Vic., NSW and southern Qld, mostly in fairly moist coastal areas.

Size 1–2 m high and about 1.5 m wide.

Soil Well-drained composted sand or light to medium loam with light mulch. Avoid root disturbance.

Climate Cooler temperate to semi-tropical, rarely happy in dry inland areas but frost tolerant.

Aspect Full to lightly broken sunlight is best. Water: Liberal to moderate.

Prostanthera linearis

Feeding Slow release, organic fertiliser or fowl manure used sparingly in spring just out from dripline of plant.
Pruning Light pruning after flowering prolongs life of plant.
Propagation Cuttings.
Possible Problems Not long lived; therefore best replaced from easily grown cuttings every four to six years. As an alternative they can be grafted on to *Westringea*.

Prostanthera lasianthos

Victoria's Christmas Bush

FAMILY LAMIACEAE

A small erect and fast growing tree or large shrub with slender lance-shaped leaves to 10 cm long, often bronze-green when in sun, with sprays of white, pink or more commonly pale mauve flowers in early to mid-summer. A good screen plant for southern States or highland areas.

Origin Tas., Vic. and mainly on highlands of NSW.
Size Shrubs to 2 m or trees pyramid-shaped with erect woody trunks to 10 m then often in a copse.
Soil Found frequently on deep volcanic or granite soils but adapts to most well-drained loams or composted sand.
Climate Cool to temperate or elevated semi-tropical; tolerates severe frost.
Aspect Grows equally well in cool shaded gullies or fairly exposed hillsides, the latter more in cooler areas.
Water Liberal but withstands long dry periods.
Feeding If at all, lightly with organic or slow release fertiliser or weathered animal manure.
Pruning May be cut back after flowering.
Propagation Cuttings.
Possible Problems Comparatively long lived and usually trouble free.

Prostanthera magnifica

Prostanthera linearis

Mint Bush

FAMILY LAMIACEAE

A fairly compact shrub with slender linear leaves about 2.5 cm long and in spring usually with blue to pale violet but sometimes mauve or white flowers.

Origin NSW and southern Qld.
Size 1–2.5 m high and 1–2 m wide.
Soil Adaptable to all but tough clay or unmulched sand providing drainage is good.
Climate Cooler temperate to semi-tropical; frost tolerant.
Aspect Lightly broken sunlight, dappled shade or direct sunlight with protection during hottest hours.
Water Liberal to moderate; tolerates long dry periods when established.
Feeding Sparingly with slow release or organic fertiliser but leafmould is usually sufficient.
Pruning Prune back to about one-third after flowering providing stems are not denuded of foliage.
Propagation Cuttings.
Possible Problems See *Prostanthera incisa*.

Prostanthera magnifica

FAMILY LAMIACEAE

An unusual shrub with oblong to lance-shaped leaves about 2 cm long and from late winter to late spring the ends of branches packed with a profusion of large flowers. These are pale blue to mauve, about 2.5 cm long and made even more striking by the large leathery reddish purple calyx protruding over each one like the upturned broad rim of a bonnet. These remain so after the flower finishes.

Origin WA, west and north-west of Geraldton.
Size Up to 2.5 m high in native habitats, usually about 1–1.5 m in cultivation and to 1 m wide; comparatively slow growing.
Soil Well-drained and no more than mildly acid composted sand or gritty loam.
Climate Temperate to tropical but preferably not exposed to hot wet summer conditions; tolerates only moderate frosts.
Aspect Lightly broken sunlight is preferable.
Water Moderate only, allowing surface soil to dry between waterings.
Feeding Very sparingly if at all with blood and bone, slow release or weathered animal manure.
Pruning Prune back old flower stems to about one-third length providing they are not denuded of foliage.
Propagation Cuttings or grafting.
Possible Problems Usually short lived in wet summer climates unless grafted on *Westringea* root stock.

Prostanthera nivea

Snowy Mint Bush
FAMILY LAMIACEAE

A lovely prostanthera with light green linear leaves to 4 cm long and during spring massed with large lacy white flowers. There is also a mauve-blue form.

Origin Vic., NSW and southern Qld.
Size 2–3 m high and as wide.
Soil Most well-drained and not too compacted garden soils.
Climate Cool to semi-tropical; moderate frost tolerance.
Aspect Will grow in full sun but best with lightly broken to dappled sunlight.
Water Liberal to moderate.
Feeding Sparingly with slow release, organic-type or fowl manure in spring.
Pruning Needs pruning back by at least one-third after flowering but without denuding stems of foliage.
Propagation Cuttings.
Possible Problems Unless grafted is fairly short lived but grows and flowers quickly therefore well worth occasional replacement.

Prostanthera ovalifolia

Mint Bush
FAMILY LAMIACEAE

The well known 'mint bush' that has enjoyed popularity in so many gardens since the late 1920s. The foliage is variable, usually 2–3 cm long, sometimes nearly oval as the species name implies but can be lance shaped or almost elliptical, always soft sage green with musky aroma. Tapering sprays of lavender-purple flowers almost cover the cone to pyramid-shaped bush in spring.

Origin Qld and NSW, particularly in sandstone areas between Dividing Range and coast.
Size 2–3 m high and to 2 m wide.
Soil Well-drained composted sand to all but the heaviest loams.
Climate Cooler temperate to semi-tropical; frost tolerant.
Aspect Generally found naturally in lightly broken sunlight but also accepts full sun.
Water Liberal to moderate.
Feeding Sparingly, in spring with slow release, complete organic fertilisers or fowl manure.
Pruning Prune back as flowers finish, at least lightly, but preferably to about one-third of the newer growth.
Propagation Cuttings.
Possible Problems As for *Prostanthera incisa* but flowers within a year of planting and well worth occasional replacement.

Prostanthera nivea

Prostanthera ovalifolia

Prostanthera x 'Poorinda Ballerina'

Prostanthera x 'Poorinda Ballerina'

FAMILY LAMIACEAE

A hybrid with close resemblance to *Prostanthera striatiflora* but it is much more compact with smaller oblong foliage and the flowers are more concentrated with a mauve flush. During full flower in spring it resembles a lacy canopy.

Origin Hybrid, probably from *Prostanthera nivea* or *P. striatiflora*.
Size Height to about 1.5 m and nearly as wide.
Soil Most well-drained garden soils.
Climate Cool to semi-tropical.
Aspect At least half sun.
Water Liberal to moderate.
Feeding Sparingly with slow release or organic type in spring.
Pruning Prune back by about one-third after flowering.
Propagation Cuttings.
Possible Problems Appears to be trouble free.

Prostanthera seiberi

Mint Bush

FAMILY LAMIACEAE

Small rounded shrub with light green oval leaves to 1.5 cm long, margins rolled under, slightly indented or toothed towards the tip giving a triangular appearance, and musky aroma. Small clusters of lavender-purple flowers are scattered over the bush in spring.

Origin NSW.
Size 1.5 m high and as wide.
Soil Well-drained, composted sand or heavy soils lightened with leafmould.
Climate Cooler temperate to semi-tropical; frost tolerant.
Aspect Happiest in lightly broken sunlight or dappled shade.
Water Liberal to moderate.
Feeding Sparingly with slow release or complete organic plant food in spring.
Pruning Lightly after flowering.
Propagation Cuttings.
Possible Problems Fairly trouble free but best replaced from cuttings after about five years.

Prostanthera striatiflora

FAMILY LAMIACEAEE

A rather open erect shrub with smooth mid-green oblong leaves, to 2.5 cm. During spring it displays very long lipped white flowers striped purplish brown on either side of the yellow stained throat.

Prostanthera striatiflora

Origin NT and inland areas of Qld, NSW, SA and WA.
Size Variable from 0.75–1.5 m high and as wide.
Soil At home in red sands of the centre and gravelly soils but accepts all but very acid or heavy clay soils providing drainage is good.
Climate Temperate to tropical; frost tolerant.
Aspect Lightly broken to full sunlight.
Water Moderate only.
Feeding Light dressing of complete, slow release or fowl manure in spring.
Pruning Lightly after flowering.
Propagation Cuttings.
Possible Problems Unless grafted may deteriorate with root rot in wet summer areas.

Prostanthera teretifolia

FAMILY LAMIACEAE

A very open shrub with fairly sparse small needle-like (terete) foliage to about 1.5 cm long and during spring attractive mauve-blue flowers, frequently in pairs in the leaf axils.

Origin Inland NSW.
Size To about 2 m high and nearly as wide.
Soil Most well-drained soils; if heavy loam, surface should be kept open with leafmould or gravelly scree.
Climate Temperate and semi-tropical.
Aspect At least half sun, preferably an open position with good air circulation.
Water Moderate only.
Feeding Sparingly with slow release, or organic fertiliser in spring.
Pruning Prune back by about one-third after flowering.
Propagation Cuttings.
Possible Problems Likely to be short lived in wet east coast areas.

Prostanthera seiberi

Prostanthera teretifolia

Ptilotus aristatus

Violet Mulla Mulla,
Pussy Tails

FAMILY AMARANTHACEAE

A perennial that is sometimes listed as *Ptilotus exaltatus* which although similar has consistently long flower stems (to 60 cm) and shorter flower spikes. *P. aristatus* has spreading basal stems, thick oval to spatulate leaves to 8 cm long, stem leaves small and tending towards lance shaped, which are smooth green when mature. It has cylindrical to conical flower spikes 10–15 cm long, violet-grey, their stems elongating as lowest flowers finish and remaining clasped to the stem. Flowering time is from late winter to early summer.

Origin Central Australia and spilling into more arid areas of surrounding States.
Size Some stems to about 50 cm, width 75 cm.
Soil Well-drained soil; found mostly in gravelly sand.
Climate Temperate to drier tropical; frost tolerant.
Aspect Open, sunny.
Water Moderate only; once established leave soil dry between occasional soakings.
Feeding Light scattering of complete fertiliser or fowl manure in spring and autumn.
Pruning Remove old flower stems.
Propagation Seed.
Possible Problems May be short lived in more humid districts.

Ptilotus aristatus

Ptilotus exaltatus

Ptilotus exaltatus

FAMILY AMARANTHACEAE

Similar to *Ptilotus aristatus* but taller. It has heads of mauve-grey to lilac flowers, initially conical to about 7 cm then elongating to cylindrical.

All other information as for *P. aristatus.*

Ptilotus macrocephalus

Green Mulla Mulla

FAMILY AMARANTHACEAE

This rather tufted species has smooth slender linear to lance-shaped leaves, to 16 cm long, and numerous stems with solitary globular spikes of initially

Ptilotus macrocephalus

fragrant silvery green flowers becoming more cylindrical as lower florets finish.

Origin Central Australia, lower more arid area of NT and arid areas of neighbouring States.
Size Stems reach 0.75–1 m high and the clump is about 0.5 m wide.
Soil Grows on dry rocky outcrops or in occasionally swampy lowlands but a well-drained soil with compost added is the safest location.

Other requirements as for *Ptilotus aristatus*.

Pultenaea daphnoides

Pultenaea

and other pea flowers

In the vast pea family (Papilionaceae or alternatively Fabeacea) to which Pultenaea belongs there are nearly 100 other genera and most of these with a number of species (Pultenaea has over 100!). To make matters more confusing, so many of them have yellow and red flowers which automatically seems to confer the same common name Eggs and Bacon.

Distinguishing between these pea flowers is not easy but here is one approach that hopefully will help to unscramble at least some of them:

Firstly the family is divided into three groups according to the formation of their stamens (all have ten stamens). In the first small group (about 10 per cent of the entire family including *Pultenaea*) all ten stamens remain separate from each other right to their base. In the remainder of the family the stamens are either all fused together until they are approaching the anthers at their outer tips, or all but one (usually the uppermost one) are fused.

Then there are other less subtle characteristics that distinguish one genus from another within the group. For example, *Pultenaea* carries all its flowers in terminal clusters, fairly evenly spaced (or more often crowded) around the tips of its stems or branchlets and more usually surrounded by a backing of leaves whereas others with similar flowers carry them differently. On *Dillwynia* they are scattered in the leaf axils towards the tips of branchlets. *Davisia* carries flowers on short racemes coming from the leaf axils and *Oxylobium* similarly but with racemes usually coming from the junction of a whorl of three leaves with stem.

Yet another easily noticed characteristic of *Pultenaea* is that with most species, stipules (short leaf spurs) are attached to the petiole (leaf stalk) base at its junction with the stem and come away with it if a leaf is stripped off the stem.

Other small characteristics helping to identify pea flowers that otherwise look similar is that *Dillwynia* has alternate leaves, the standard (large erect petal at top of flower) is broader than long. This also applies to *Eutaxia* but leaves are opposite. Then there are differences in calyx lobes, seed pods etc. Even different chromosome patterns are now used to help in identification.

Pultenaea daphnoides

Eggs and Bacon

FAMILY PAPILIONACEAE

A usually tall slender shrub with leaves very dark green above, wedge shaped (obovate), about 2 cm long, and during spring terminal clusters of pale orange pea flowers with a dull red centre. The species name (derivation 'like a daphne') alludes to *Daphne collina*, not the better known *D. odora* which has much larger leaves.

Origin All States except WA, especially in Hawkesbury areas; also in New Zealand.
Size 2–3 m high, 1–1.5 m wide.
Soil Very much at home in well-composted sandstone areas but adapts to most well-drained garden soils.
Climate Cooler temperate to semi-tropical; frost tolerant.
Aspect Best in broken sunlight to dappled shade.
Water Liberal to moderate; good drought tolerance.
Feeding Leafmould is sufficient, otherwise very light applications of slow release or complete organic food in spring.
Pruning Lightly after flowering. Severe pruning into leafless wood can kill this plant. Tip prune young growth.
Propagation Scarified seed.
Possible Problems Rarely a long lived shrub but quick to flower.

Pultenaea flexilis

Pultenaea flexilis

Eggs and Bacon

FAMILY PAPILIONACEAE

A spreading medium to large shrub with gracefully pendulous outer branches and narrow oblong leaves to 2 cm. In spring the outer branches are loaded with flowers, making a delightful cascade of deep golden yellow, especially in lightly shaded moist coastal gullies.

Origin Qld and NSW, mainly from Dividing Range to coast.
Size 3-4 cm high and about 2-2.5 cm across; smaller in full sun.
Soil Grows well in composted sand or heavier loams with a leafy mulch providing drainage is good.
Climate Temperate to semi-tropical.
Aspect Best in dappled shade to about half sun.
Water Liberal to moderate; some drought tolerance when established.
Feeding Should not be necessary if leafmould is well supplied. If any, organic or slow release fertiliser.
Pruning Prune outer branches back to about half after flowering.
Propagation Scarified seed.
Possible Problems Usually trouble free.

Pultenaea pedunculata

Pultenaea procumbens

Pultenaea pedunculata

Matted Bush Pea

FAMILY PAPILIONACEAE

An attractive little plant for rock gardens with prostrate selflayering stems, densely clustered slender dark green leaves about 0.5 cm long and orange, red and yellow or clear bright yellow pea flowers in small clusters on tiny stems from the leaf axils. 'Pyalong Gold' (illustrated) is a bright yellow form growing at the National Botanic Gardens, Canberra.

Origin SA, Tas., Vic. and NSW.
Size To 10 cm high and spreading about 1.5–2 m.
Soil Well-drained sandy loam with compost added or clayey loam lightened with coarse sand and leafmould.
Climate Cool and temperate; frost and sea spray tolerance.
Aspect At least half sun or lightly broken sunlight and avoid crowding by other plants.
Water Moderate to liberal; accepts long dry periods.
Feeding Sparingly with organic plant food in spring.
Pruning Rarely necessary.
Propagation Scarified seed.
Possible Problems May be subject to fungal rot in hot wet summer areas; growing on a surface scree should help overcome this.

Pultenaea procumbens

Scrambled Eggs and Bacon

FAMILY PAPILIONACEAE

An interesting trailing plant for rock gardens, etc., with contour-following branches with dark green hairy slender pointed leaves and a cascade of small pale orange and brown pea flowers during spring.

Origin Vic. and NSW, tableland areas.
Size 0.3 cm high with a spread of 1–1.5 m.
Soil Needs well-drained and preferably composted light loam.
Climate Cool and temperate; frost tolerant.
Aspect Dappled sunlight or morning sun only.,
Water Liberal to moderate; tolerates fairly long dry periods.
Feeding Very sparingly; no more than light application of organic plant food in spring.
Pruning Only if necessary to thicken and remove twiggy growth.
Propagation Scarified seed.
Possible Problems So far no recorded difficulties.

Pultenaea stipularis

Eggs and Bacon

FAMILY PAPILIONACEAE

Low spreading with fairly rigid erect stems densely clad in pine-like leaves to 4 cm long and, during spring, crowded terminal heads of yellow pea flowers.

Origin NSW.
Size About 1 m high and as wide; more open and narrower in shade.
Soil Adapts to most well-drained garden soils with a preference for sandy loam with leafy mulch.
Climate Cool temperate to semi-tropical; frost tolerance.
Aspect Full sun to light shade.
Water Liberal to moderate; accepts dry periods when established.
Feeding Sparingly with slowly soluble or organic fertiliser in spring. Seaweed extract is also helpful.

Pultenaea villosa

Pultenaea stipularis

Pultenaea weindorferi

Pruning Stems may be pruned back to about half after flowering providing some lower foliage is retained.
Propagation Scarified seed.
Possible Problems Apart from an occasional tip-webbing caterpillar is usually trouble free.

Pultenaea villosa

Eggs and Bacon

FAMILY PAPILIONACEAE

A delightful trailer or prostrate shrub with grey-green hairy narrow oblong leaves about 0.5 cm long and a profusion of small flower clusters on tiny branchlets during spring. Makes an eye-catching golden cascade over rockeries, walls or from elevated containers.

Origin Qld and NSW.
Size Main stem can be trained erect and branches trailing or allowed to follow contours. Spreads to about 2 m.
Soil Most well-drained soils, preferably with leafmould or compost added.
Climate Cool to semi-tropical; frost tolerant.
Aspect Lightly broken sunlight to dappled shade; also accepts full sun.
Water Liberal to moderate.
Feeding Sparingly with complete organic or slowly soluble fertiliser.
Pruning Lightly prune back branches, especially any ascending ones after flowering.
Propagation Scarified seed.
Possible Problems Usually trouble free.

Pultenaea weindorferi

Swamp Bush Pea

FAMILY PAPILIONACEAE

A small prostrate to slightly ascending shrub with some resemblance to *Pultenaea stipularis*, having similar but smaller pine-like leaves to 1 cm long and dense terminal clusters of golden yellow pea flowers with a very small red centre blotch during spring.

Origin Vic.
Size Several forms, from 0.5–1.5 m high and spreading up to 2 m.
Soil Moist but well-drained composted sand to medium loam.
Climate Cool to temperate; good frost tolerance.
Aspect Sun to light shade.
Water Liberal; resents long periods of dryness.
Feeding Sparingly with organic plant food or small amount of slow release fertiliser in spring.
Pruning Ascending stems may be cut back heavily after flowering.
Propagation Scarified seed.
Possible Problems Apart from an occasional tip-webbing caterpillar seems trouble free.

Rhododendron lochae

Rhododendron lochae

FAMILY ERICACEAE

Australia's only native rhododendron. A small shrub with dark glossy green and thick textured oval leaves 5–7 cm long and loose terminal clusters of bright coral-red waxy trumpet-shaped flowers flaring to a mouth about 5 cm across. These occur mainly during late summer and early autumn.

Origin Northern Qld, on elevated peaks of Cape York.
Size 75 cm high and similar width.
Soil They are best in a container, or in a pocket in the garden with a soil coarsened by the addition of rotted leafmould, weathered chopped bark and peatmoss or even a cymbidium orchid mix; one that holds moisture but drains quickly.
Climate Temperate to tropical, preferably humid; accepts temperatures down to -3°C for brief periods only.
Aspect Lightly broken sunlight to dappled shade; more sun is acceptable in areas where humidity is very high.
Water Needs to be kept well watered during summer.
Feeding Complete water soluble fertilisers may be used monthly from late spring to late summer.
Pruning If plant is becoming very leggy,

try pruning by degrees (branch at a time) from mid spring to late summer, even at the expense of flower.
Propagation Layering, cuttings or seed.
Possible Problems Thrips or red spider mite may mottle or silver foliage, especially of plants under cover.

Ricinocarpos bowmanii

FAMILY EUPHORBIACEAE

Small shrub with very slender linear to oblong leaves about 2 cm long which is spangled with flowers about 2.5 cm across, the shape of a five-pointed star, white to blush pink with a deeper pink centre. Flowers appear during spring.

Origin Qld and NSW.
Size To 1 m high and at least as wide.
Climate Cooler temperate to semi-tropical; frost tolerant.
Aspect Lightly broken sunlight or about half sun seems best but they also accept full sun.
Water Moderate to liberal; seems to accept at least brief dry periods.
Feeding Light application of slow release or organic fertiliser in spring is acceptable; otherwise a leafy mulch seems adequate.
Pruning Light to moderate after flowering otherwise growth becomes twiggy.
Propagation Cuttings or seed.
Possible Problems Seems to be trouble free.

Ricinocarpos pinifolius

Wedding Bush

FAMILY EUPHORBIACEAE

An erect small shrub with very narrow stiff dark green linear leaves, to 4 cm long, downward rolled at margins, and suggesting a lacy veil in spring when massed in white flowers about 2.5 cm across with five and occasionally six slender petals.

Origin All eastern States including Tas., particularly on north-eastern sandstone hillsides and coastal moorland.
Size 1 m high and nearly as wide.
Soil Prefers composted sand or light loam but adapts to all but the heaviest clays if drainage is good.
Climate Cool to semi-tropical; tolerant to frost and salt winds.
Aspect Best with lightly broken sunlight but also grows in full sun.
Water Liberal to moderate; withstands dry periods.
Feeding Accepts a light sprinkling of fowl manure, organic or slow release fertiliser in spring.
Pruning May be lightly pruned after flowering.
Propagation Cuttings.
Possible Problems Usually trouble free.

Ricinocarpos bowmanii

Ricinocarpos pinifolius

Rulingia hermanniifolia

Dwarf Kerrawang

FAMILY STERULIACEAE

A prostrate groundcover shrub with a rigid mat of spreading branches, stemless (sessile) leathery lance-shaped to oblong leaves to 2 cm long, margins downward rolled, and, during spring, almost covered in small terminal heads of tiny pinkish white starry flowers. These are followed by bristle-covered brown capsules.

Origin NSW.
Size To 30 cm high and to about 1.5 m wide.
Soil Most well-drained garden soils.
Climate Cool temperate to semi-tropical; frost tolerant.
Aspect Full sun or light shade.
Water Once established moderate only. Allow surface to dry between waterings.
Feeding Little needed; optional light scattering of organic or slow release fertiliser in spring.
Pruning Only for shape or shortening back to thicken branches.
Propagation Cuttings.
Possible Problems Seems trouble free.

Rulingia hermanniifolia

Scaevola aemula

Fairy Fan Flower
FAMILY GOODENIACEAE

A very variable plant according to area of origin. All are attractive close matting trailers. The more popular prostrate form has fairly thick bright green wedge-shaped leaves to 5 cm long, some with margins lightly serrated towards the tip, and pale mauve to purple five-lobed flowers, some forms about 1 cm others to 2 cm across.

Scaevola aemula

Origin Eastern Australia; other forms are from WA, inland areas of mainland States and Tas. They flower in spring and summer.
Size Mats 10–12 cm high and spreading about 1 m.
Soil All prefer a well-drained soil lightened if necessary with leafmould or coarse sand.
Climate Cool to semi-tropical; tolerant to salt spray and most possibly to frost.
Aspect Open sunny to dappled sunlight.
Water Liberal to moderate but tolerant to fairly long dry periods.
Feeding Light dressing of organic fertiliser or even slow release; can be spread and washed off foliage, in spring.
Pruning Mainly to control size and shape; pinching back tips of runners induces branching.
Propagation Cuttings or with care from layers.
Possible Problems All are easy to maintain except perhaps a suckering form from western NSW which needs excellent drainage.

Scaevola calendulaceae

Dune Fan Flower
FAMILY GOODENIACEAE

Prostrate growth similar to *Scaevela aemula*, glossy, stalked (petioled) leaves to about 7 cm, mostly smooth edged, and spikes of blue or blue and white flowers on erect spikes.

Origin Coastal areas of eastern Australia and SA, often on sand dunes.
Size Mat as *S. aemula*, flower spikes may be 15–20 cm.

Requirements as for *S. aemula*.

Scaevola ramosissima

Fan Flower
FAMILY GOODENIACEAE

A sprawling plant with glossy green, but hairy, wandering stems, trailing or sometimes climbing to about 1 m if convenient support from other plants is available. Leaves are lance shaped, 2–4 cm long. Scattered flowers are 2–2.5 cm across, deep purple to soft lavender blue with a yellow and white throat; they appear mainly from late winter into summer but sporadic.

Origin Eastern Australia, normally in lightly timbered areas.
Size Without support 10–12 cm high, spreading 1–1.5 m.
Soil More common in sandstone well mulched with leafmould but adapts to most well-drained soils.
Climate Cooler temperate to semi-tropical; tolerates salt winds; resents severe frost but can be growned within shelter of other shrubbery.
Aspect Part sun or lightly broken sunlight to dappled shade.
Water Liberal to moderate; good drought resistance when established.
Feeding Light sprinkle of organic manure in spring but leafmould is generally sufficient.
Pruning Stems may be tip pruned to encourage branching.
Propagation Cuttings, layering or seed.
Possible Problems Occasion leaf-rolling caterpillar.

Scaevola calendulaceae

Scleranthus biflorus

Canberra Grass

FAMILY CARYOPHYLLACEAE

This plant makes low attractive mounds of what looks like emerald green moss. It looks most effective in the foreground of rock gardens, between stepping stones, etc. It produces tiny white flowers but is grown for colour, foliage, texture and form which suggests cool moist shade but it needs sun to do well.

Scaevola ramosissima

Scleranthus biflorus

Senecio magnificus

Sowerbaea laxifolia

Origin Tablelands and several inland areas of eastern Australian States.
Size Mounds reach about 8 cm high and 40–50 cm across.
Soil Any average garden soil.
Climate Cool to semi-tropical; frost tolerant.
Aspect Needs almost full sun to retain good compact form.
Water Liberal.
Feeding Should not be necessary; if soil is very impoverished it may have one or two applications of complete water soluble fertiliser mixed to half strength; frequent feeding spoils form.
Pruning Not needed unless necessary to check spread.
Propagation Dividing clumps during cooler months or seed.
Possible Problems A fungus disease sometimes causes dead patches. These can be cut out and clump sprayed thoroughly with Benlate. Recovery is fairly rapid.

Senecio magnificus

Tall Yellow Top

FAMILY ASTERACEAE

A showy perennial with blue-grey fleshy oblong to lance-shaped stem-clasping leaves 4–8 cm long, sometimes toothed, and large branching heads of bright yellow daisies to about 3 cm across throughout the year but most profuse after rains, especially in spring.

Origin Occurs in all States except Tas. usually in more arid areas along roadsides, in drains where water collects after rain. Profuse in southern half of NT.
Size 0.6–1 m high and nearly as wide.
Soil Preference is for well-drained slightly alkaline loam.
Climate Temperate and semi-tropical; frost tolerant.
Aspect Open sunny position.
Water Moderate, allowing surface soil to dry for a few centimetres between soakings.
Feeding Light dressing of complete fertiliser, mainly spring and autumn followed by a good soaking.
Pruning Remove old flower stems.
Propagation Cuttings or seed.
Possible Problems Reputed to have caused stock poisoning.

Sowerbaea laxifolia

Vanilla Lily

FAMILY LILIACEAE

Similar to *Sowerbaea juncea* but a little larger and more robust both in growth and flowers, these with longer and broader petals tending more to purple in colour than violet.

Origin WA, mainly in Perth area and west of Darling Range.
Size As above.
Soil At home in sandy to gravelly soils and because of its location, obviously accepts drier conditions during summer.

Other requirements as for *S. juncea*.

Sollya heterophylla

Sollya heterophylla

(syn *Sollya fusiformis*)

Blue Bell Creeper

FAMILY PITTOSPORACEAE

This twining plant can be used as a gentle climber or fence cover or there is a form more suitable for a low mounding shrub. Stems are brown and wire like, leaves bright green, glossy slender oblong to lance shape, 3–5 cm long. During late spring and summer slender bell-shaped flowers 1.5–2 cm long, usually sky blue but there is also a pale blue and occasionally a pink form. These are followed by dark bluish cylindrical fruit or seed pods.

Origin WA, mainly in the Perth area and south.
Size Climbing to about 2 m in all directions; as a shrub can be a dense mound about 1 m high or can be allowed to mat as a groundcover. Effectively used in the latter style in Tel Aviv and other Mediterranean areas.
Soil Adapts to most well-drained soils.
Climate Cooler temperate to semi-tropical; tolerates moderate frosts.
Aspect Full sun to light shade.
Water Liberal to moderate; grows better with moisture but accepts drought periods.
Feeding Slow release or organic fertiliser in spring or early summer.
Pruning Only if necessary or to encourage branching.
Propagation Cuttings and seed.
Possible Problems Usually trouble free in most districts.

Sprengelia incarnata

Stackhousia monogyna

Sprengelia incarnata

Swamp Heath

FAMILY ERICACEAE

An erect plant with stems and outer branches almost completely covered with closely set, broad based, but sharply tapering and pointed, dark glossy green, stem-clasping leaves only about 1 cm long. These lead up to dense terminal cone-shaped clusters of pink-flushed porcelain-like flowers with finely tapering reflexed petals and intermingling bright pink buds. They appear mainly during spring and summer. A useful and interesting plant for damp areas and an excellent cut flower.

Origin SA, Tas., Vic. and NSW, mainly in damp peaty areas near swamps or in mossy areas where seepage is continuously present.
Size 1–2 m high and only about 50 cm wide; for best effect plant several in a clump.
Soil Well-composted moist, but not stagnant, peaty soils; no lime.
Climate Cool and temperate.
Aspect Lightly broken to full sunlight.
Water Liberal; should not completely dry out.
Feeding Sensitive to strong fertilisers; azalea food, or animal manure fairly well decomposed may be used in spring but like many other plants they can survive well on leafy mulches alone.
Pruning More compact with more flower stems if pruned back by at least one-third after flowering.
Propagation Cuttings.
Possible Problems Usually trouble free if soil is moist or slightly acid.

Stackhousia monogyna

FAMILY STACKHOUSIACEAE

A clumpy perennial, stems and slender 1–2 cm leaves are glossy bright green. Long spikes of starry white or cream flowers appear during spring or early summer.

Origin Tas., SA, Vic., NSW and Qld, on tablelands and into some Mallee areas.
Size To 50 cm high and spreading to nearly 1 m.
Soil Best in gritty coarse sandy loam,

volcanic or other quick draining soil, elevated rocky pockets, etc.
Climate Cold and temperate.
Aspect From lightly broken sunlight to light shade.
Water Liberal; although good drainage is needed it resents any but brief dryness.
Feeding Light sprinkling of complete fertiliser, slow release or fowl manure in autumn and early spring.
Pruning Cut back old flower stems.
Propagation Base cuttings from crown.
Possible Problems Fairly trouble free if even moisture is combined with sharp drainage.

Stenocarpus sinuatus

Qld Fire Wheel,
Wheel Tree
FAMILY PROTEACEAE

A beautiful erect tree with dark glossy leaves to about 30 cm long, some slender oblong, majority irregularly lobed, and large showy clusters of red flowers starting like the spokes and hub of a wheel when in tight brownish green buds then gradually turning red as they expand to 6–7 cm in diameter before the long gold-tipped stamens separate and give them a spidery appearance, mainly in late summer and autumn. These are followed by pendulous clusters of slender seed pods. A tree often bypassed because of its slow-to-flower reputation (five to seven years) but looks attractive even in early growth and is well worth waiting for.

Origin Qld and northern NSW, rainforest areas.
Size 6–10 m high and 3–4 m wide; much taller in more natural habitats.
Soil Very adaptable even accepting and making good growth in heavy moist clay if started with the addition of improved soil but reasonable drainage is preferable.
Climate Moderately cool temperate to tropical; needs protection from heavy frosts for the first few winters.
Aspect May be started in full sun, light shade or under the protection of shrubbery to grow into full sun.
Water Liberal but withstands dry periods when established.
Feeding Water soluble or slow release fertilisers in spring and summer once established to speed early growth, plus a leafy mulch.
Pruning Only if needed for shape.
Propagation Seed.
Possible Problems Macadamia twig girdler but rarely causes serious problems.

Stenocarpus sinuatus

Stylidium graminifolium

Trigger Plant
FAMILY STYLIDIACEAE

This species makes a rosette of rather wiry and slender foliage varying from 8–18 cm long according to environment. Flower stems are also slender and erect, terminating in a spike carrying from ten to fifty flower buds gradually opening from the base, colour varying from pale pink to rosy purple. The unique character about this genus is that pollen and female stigma is mounted on a sensitive spring like column or 'trigger' held at the 'ready' position nearly 1 cm out from the flower centre. When agitated by a nectar-seeking insect, this trigger springs down across the flower, depositing pollen on the insects back and hopefully at the same time gathering more from another flower on its stigma. The flower is capable of 'reloading' during warm sunny periods.

Origin SA and all eastern States, mainly from coast and mountains including snow regions.
Size 20–40 cm high including flower stem, clump spread to about 15 cm.
Soil Adapts to a wide range of preferably well-drained and composted soils.
Climate Cold to semi-tropical; frost and snow tolerant.
Aspect Lightly broken to full sunlight.
Water Liberal to moderate.
Feeding Leafmould is sufficient, or sparingly with either slow release or organic fertilisers.
Pruning Remove old flower spikes.
Propagation Seed.
Possible Problems Usually trouble free.

Stylidium bulbiferum

Trigger Plant
FAMILY STYLIDIACEAE

This interesting plant expands by stolons making low clumps of dark glossy green grass-like foliage and in spring it is almost covered with quaint triggered flowers, usually pink but may

Stylidium graminifolium

be deep rosy red or white; makes an interesting container plant. For detail of flowers see *Stylidium graminifolium.*

Origin WA, coastal areas.
Size Circular mats 12–15 cm high and about 30 cm across.
Soil Quick draining gritty soil or medium loam with up to 50 per cent coarse river sand and peatmoss or rotted leafmould added.
Climate Temperate to semi-tropical; very little frost tolerance.
Aspect Lightly broken to full sunlight.
Water Liberal, but well drained, such as where seepage water gets away well.
Feeding Light dressing with organic fertiliser in spring or autumn.
Pruning Remove old flower stems.
Propagation By division of clumps or seed.
Possible Problems Usually trouble free.

Styphelia triflora

Five Corners

FAMILY EPACRIDACEAE

An erect shrub with rigid and closely set blue-green lance-shaped leaves 3–5 cm

Stylidium bulbiferum

Styphelia triflora

Styphelia tubiflora

Syzygium luehmannii

long and beautiful clusters of pink tubular flowers with five long hairy pale yellow lobes reflexed and rolled back at the tip above long protruding pinkish stamens. These delightful flowers appear in spring. There is also a green-flowered form in ACT (apart from green-flowered species *Styphelia laeta*).

Origin NSW, especially from coast to ranges.
Size To 1.5 m high and occasionally as wide (usually about 0.75 m).
Soil Well-drained and composted sand to light loam and a leafy mulch.
Climate Cooler temperate to semi-tropical; frost tolerant.
Aspect Lightly broken sunlight to full sun.
Water Liberal to moderate once established; accepts dry periods.
Feeding Very sparingly if at all, with organic or slow release fertiliser in spring.
Pruning Prune back moderately after flowering, if possible removing old flower clusters without sacrificing too much new growth. Tip prune when young to encourage branching.
Propagation Cuttings.
Possible Problems Occasional tip-rolling caterpillars, but they do not cause serious damage.

Styphelia tubiflora

FAMILY EPACRIDACEAE

Small meandering shrub suitable for a rockery plant, with deep green 1.5 cm long, oblong or lance-shaped leaves gradually tapering to a point, and small clusters of slender tubular flowers about 3 cm long, pinkish at the calyx and deepening to cherry red at the reflexed tip.

Origin NSW, mainly central to southern coast.
Size 0.5–0.7 m high and a little wider.
Soil These occur more on the slower drying but reasonably drained medium loams than sand but the latter can be improved by adding peatmoss and in all cases compost.
Climate Temperate to semi-tropical; frost tolerant.
Aspect Lightly broken or dappled to half sunlight.
Water Liberal but usually survives dry periods if well mulched.
Feeding Sparingly if at all with organic or slow release fertiliser in spring.

Pruning Prune back by about one-third after flowering, providing branches are not denuded of foliage.
Propagation Cuttings.
Possible Problems Usually trouble free if not overfed or left thirsty for too long.

Syzygium luehmannii

(syn. *Eugenia luehmannii*)

Small Leafed Lillypilly

FAMILY MYRTACEAE

A small to medium sized tree with attractive dense foliage and fruit, glossy green lance-shaped leaves to 6 cm, pink to coppery red when young. White fluffy spring flowers are followed by small pendulous pear-shaped berries turning coral red when mature.

Origin Qld and NSW, mainly in moist gullies.
Size Dense low branching to 5 m for first ten years then eventually to 10 m and 5–6 m wide.
Soil Adapts to most well-drained and composted garden soils.
Climate Temperate to tropical; not recommended for severe frost areas but grows in most parts of Melbourne.
Aspect Open sunny position.
Water Responds best to liberal watering during summer.
Feeding Fowl manure, complete or slow release fertiliser used moderately when established.
Pruning Only for shape if necessary.
Propagation Seed.
Possible Problems Less affected by scale and leaf blister than most other lillypillies.

Tasmannia purpurascens

(syn. *Drimys purpurascens*)

Pepper Bush

FAMILY WINTERACEAE

Medium sized shrub with reddish brown stems, purple-bronze new leaves normally maturing dark green, leathery wedge shaped, to about 15 cm long, and in late spring to summer terminal clusters of flowers with prominent cream stamens arranged in globular form and backed by thin cream petals. Black berries with peppery flavour follow in late summer–autumn.

Origin NSW, Barrington Tops behind lower north coast.
Size 2 m high and about 1.5 m wide.
Soil Well-drained and composted sandy to heavy loam.
Climate Cooler temperate to semi-tropical; frost tolerant.
Aspect Dappled or about half sunlight, preferably with some shelter from other growth.
Water Liberal to moderate.
Feeding Light dressing of complete organic or slow release fertiliser in spring when plants establish.
Pruning May be lightly pruned if necessary, preferably before new growth (before flowers finish).
Propagation Seeds or cuttings.
Possible Problems Usually trouble free. Bleached margins of illustrated plant was caused by overfeeding.

Tasmannia purpurascens

Telopea

It is suggested that a mycorrhiza (friendly fungus or bacteria) needs to be introduced from soils where telopea grow successfully or even from banksia leafmould. However, years of research has not produced a conclusive answer at the time of writing. One obvious point is that growth and performance does improve for at least a few years after a bushfire.

One theory is that the fire eliminates retarding mites, insects or other organisms. It probably also releases nitrogen from heat-destroyed protozoa, potash, calcium, etc., and in any case makes these essential inorganic substances more available by changing soil pH. in this case reducing soil acidity. It has certainly been noticed that in some bushland areas where telopeas once flourished forty to fifty years ago when occasional fires were last experienced, these plants are still alive but have diminished in size and vigour and have not flowered for many years. The soil is now highly acid (about pH4).

On the other hand one nurseryman growing telopea blooms commercially in light sandy soil, merely cuts back the plants within about 20 cm of the ground each year as flowers are finishing. He then feeds liberally with fowl manure and waters well, with great response. However, this feeding should not be tried until plants are established. In any case these unique plants are certainly worth growing, so at this stage just try and hope.

Telopea x 'Braidwood Brilliant'

FAMILY PROTEACEAE

A cross between *Telopea mongaensis* (from near Monga, NSW) and *T. speciosissima*. It makes a vigorous, more spreading and branched plant than *T. speciosissima* although there is an inconsistency in flower form, some with the more regular conical to dome-shaped head of closely packed stamens, others loose and open like *T. mongaensis*.

Telopea x 'Braidwood Brilliant'

Origin Hybrid.
Size 2–3 m high (the taller in shade) and to 3 m wide.
Soil Well-composted slightly to moderately acid soil with good drainage.
Climate Cool to semi-tropical; good frost tolerance.
Aspect Sun or shade.
Pruning Prune back.

Other requirements as for *T. speciosissima*.

Telopea speciosissima

NSW Waratah

FAMILY PROTEACEAE

The floral emblem of NSW. It is an erect shrub with stiff, wedge-shaped and usually toothed, and occasionally lobed, dark green leaves to 15 cm long and branchlets terminating in bright crimson florets densely packed into conical or peaked dome-shaped heads to 15 cm across, and surrounded by smooth red bracts.

Origin NSW, mainly coast and tablelands.
Size 3 m high and about 1.5 m across but some branching old plants happy in cultivation are about twice this size.
Soil Well-drained and composted moderately but not highly acid, sandy to heavy loam.
Climate Cold to semi-tropical; frost resistant.
Aspect Dappled shade to full sunlight.
Water Preferably liberal during spring and early summer when new growth develops but once established withstand dry periods.
Feeding Light to moderate application of fowl manure, blood and bone, other organic or slow release fertiliser in spring to establish plants only and followed by liberal watering. Maintain leafy mulch.
Pruning Pruning back to just above last few leaves on previous season's growth (old flower stem) promotes next season's flowering, although seed pods are attractive. Pruning is preferable before new growth starts which is often before flowers lose all colour. Pruning to within about 20 cm of ground is occasionally done to successfully rejuvenate old plants.
Propagation Seed or cuttings.
Possible Problems Stem or flower bud borer. Also some plants fail to progress or old ones stagnate.

See also genus notes.

Telopea speciosissima var. 'Wirrimbirra White'

White Waratah

FAMILY PROTEACEAE

A unique white-flowered form with creamy lime bracts at the National Botanic Gardens, Canberra, and at the time of writing, difficult to obtain and rare in cultivation.

Size and cultivation details as for *Telopea speciosissima*.

Telopea truncata

Tas. Waratah

FAMILY PROTEACEAE

An erect shrub with more oblong and slightly longer leaves than those of *Telopea speciosissima* and terminal clusters of more loosely packed florets.

Origin Tas.
Size 3–4 m high and nearly as wide.

Cultivation as for *T. speciosissima*.

Telopa oreades the Gippsland Waratah, is similar to *T. truncata*, sometimes a little taller still, and with leaves to about 18 cm long.

Templetonia retusa

Bullock Bush

FAMILY PAPILIONACEAE

A spreading shrub with pleasing smooth bluish green foliage, wedge shaped to about 4 cm long with a flattened or indented apex, and large brownish to orange-red pea flowers 2–3 cm long, with an unusually large reflexed standard (upper petal), appearing during winter and early spring.

Origin SA, WA, north and south of Perth in limestone areas.
Size To 2 m high and nearly as wide.
Soil Well-drained and composted light to medium loam, preferably limey or no more than slightly acid. Addition of limestone chip would be beneficial.
Climate Cooler temperate to semi-tropical; frost tolerant.
Aspect Best in an open sunny position but dappled sunlight is acceptable.
Water Once established moderate only.
Feeding Slow release or organic plant food in autumn and spring, applied sparingly when soil is moist.
Pruning Prune after flowering.
Propagation Scarified seed.
Possible Problems Seems trouble free with soil suggested.

Telopea speciosissima

Telopea speciosissima var. 'Wirrimbirra White'

Telopea truncata

Templetonia retusa

Tetratheca bauerifolia

Black-eyed Susan,
Pink Bells

FAMILY TREMANDRACEAE

A small relatively erect many stemmed plant with star-like whorls of leaves resembling those of *Bauera* but smaller and mostly in four's and five's. Sprays of four oval to oblong rose-pink flowers 1–1.5 cm long hang like bells, hiding their black stamen cluster. This gives them the common name Black-eyed Susan. Ideal rockery plants or for underplanting lightly foliaged or high branching shrubs.

Origin Vic. and NSW.
Size 0.3–0.4 m high and nearly as wide.
Soil Composted sand or light loam; good drainage needed.
Climate Cooler temperate to semi-tropical; moderate frost tolerance.
Aspect Best in lightly broken sunlight to dappled shade.
Water Liberal to moderate; tolerates dry periods when established. A leafy mulch helps to keep them happy.
Feeding Light application of slow release or organic fertiliser in spring or just a leafy mulch.
Pruning Prune lightly after flowering.
Propagation Cuttings.
Possible Problems Usually trouble free.

Tetratheca ciliata

FAMILY TREMANDRACEAE

Similar to *Tetratheca bauerifolia* but has finer foliage, is a little more spreading

Tetratheca bauerifolia

Tetratheca ciliata

and has a deeper flower colour, tending towards rosy purple. It is also a very desirable small plant for garden or container and is more adaptable than *T. bauerifolia*.

Origin SA, Tas., Vic. and NSW.
Size 0.3–0.6 m high and to 0.6 m wide.
Soil Some preference for composted sand but also succeeds in clayey soils if rotted leafmould, coarse sand or similar lighteners are added.

Other requirements as for *T. bauerifolia*.

Tetratheca shirlessii

Black-eyed Susan
FAMILY TREMANDRACEAE

Small plant with stems appearing almost leafless because of the widely spaced nodes and often it is only younger stems and tips of older ones that retain the very small foliage. Flowers slightly smaller and deeper colour than the species previously mentioned, also a little taller.

Origin NSW, coastal areas, especially Hawkesbury sandstone country.
Size To about 0.6 m and as wide.

Requirements as for *Tetratheca bauerifolia*.

Tetratheca shirlessii

Tetratheca thymifolia

Black-eyed Susan
FAMILY TREMANDRACEAE

A popular black-eyed susan that performs well in most areas. The whorls

Tetratheca thymifolia

Tristania neriifolia

Tristania neriifolia

of tiny leaves resemble those of thyme as the species name implies. Flowers vary in size but are usually large, bright rosy mauve and freely produced. This species spreads by suckering, therefore eventually clumps are comparatively large. A white form is also known.

Origin Vic., NSW and Qld.
Size 0.3–0.6 m high and 0.6–1 m wide.
Soil Like *Tetratheca ciliata*, the preference is for a light composted soil but can be grown in heavy soils if the clay is lightened with rotted leafmould, coarse sand, etc.

Other requirements as for *T. bauerafolia*.

Tristania neriifolia

Water Gum
FAMILY MYRTACEAE

Can be grown as a dense shrub by early pruning or if in an exposed position as a small umbrella-like tree for shade. The dark green lance-shaped leaves are to 9 cm long and resembling oleander leaves, attractively surround the clusters of small yellow flowers that appear in summer.

Origin NSW, along the edges of rocky coastal creek beds.
Size 3 m high and 1–2.5 m wide, occasionally taller but controllable.
Soil Prefers damp but reasonably well-drained loam but adapts to most garden soils if well composted.
Climate Cooler temperate to semi-tropical; tolerates moderate frosts.
Aspect Sun or shade. Growth best with some shelter from strong winds.
Water Liberal but withstands dryness when established.
Feeding Most complete type fertilisers or animal manures applied moderately in spring will speed growth; providing the plant is established and soil moist.
Pruning For shape only; tip prune several times when young if low branching is preferred.
Propagation Seed.
Possible Problems Usually trouble free.

Thomasia glutinosa

FAMILY STERCULIACEAE

A small rounded shrub attractive as a rockery or container plant with small slightly glossy broad heart-shaped leaves to 2.5 cm long, often sticky. Flowers are hooded by decorative lilac-mauve calyx bracts, 2.5 cm long, and hang in pendulous clusters during spring.

Origin WA.
Size 50–70 cm high and as wide.
Soil Prefer well-composted sand or light loam with reasonable drainage. Heavier loams should be improved by the addition of coarse sand and rotted leafmould.
Climate Cooler temperate to semi-tropical; frost tolerant.
Aspect Full sun to moderate shade; an advantage is that they usually succeed in the dry shade of larger shrubs.
Water Moderate.

Feeding Very light application of organic or slow release fertiliser in autumn and spring.
Pruning Prune back lightly after flowering; also remove thin twiggy growth.
Propagation Cuttings or seed.
Possible Problems Usually trouble free.

Thomasia glutinosa

Thomasia paucifolia

FAMILY STERCULIACEAE

Small shrub with lance- to slender heart-shaped leaves 4–5 cm long and sprays of small pendulous deep lilac to pale purple flowers or bracts during spring.

Origin WA.
Size 1–1.5 m high and as wide.
Soil As for *Thomasia glutinosa.*
Climate Cool temperate to semi-tropical; frost tolerant.
Aspect Accepts full shade or sun.

Other requirements as for *T. glutinosa.*

Thomasia grandiflora

Thomasia grandiflora

FAMILY STERCULIACEAE

Small shrub with slender oblong to lance-shaped and very crinkled bright green leaves about 4 cm long and comparatively large star-shaped rosy purple- to cyclamen-coloured flowers, 2–2.5 cm across, in spring.

Origin WA.
Size 0.3–0.6 m high and 0.6–1 m wide.

Requirements as for *Thomasia glutinosa.*

Thomasia paucifolia

Thryptomene australis

FAMILY MYRTACEAE

An erect shrub with tiny leaves towards the ends of branches and very beautiful from mid-winter to late spring when the numerous short stalks towards the branch ends are heavily clustered in small pink flowers about 8 mm across with deeper pink centres. Some forms are nearly white. *Thryptomene mucronulata* is similar but with rounder leaves and is a little later flowering.

Origin WA, mainly north and north-west of Perth.

Size About 1.5 m high and nearly as wide.
Soil Well-drained composted sand or medium loam lightened with a leafy mulch; tolerates slight limeyness.
Climate Temperate; tolerates moderate frost.
Aspect Open sunny position.
Water Moderate; accepts dryness of low rainfall areas when established.
Feeding Sparingly with slow release or complete organic fertiliser in autumn and spring providing soil is moist.
Pruning Improved by pruning back moderately after flowering.
Propagation Cuttings.
Possible Problems Usually trouble free.

Thryptomene saxicola

Rock Thryptomene

FAMILY MYRTACEAE

A long flowered species, excellent for cutting and an adaptable shrub. There are several forms, those widely sold are low growing, spreading with pendulous branches which are crowded, slender oval to oblong leaves to about 7 mm. The small slightly cupped pink flowers are massed along outer branchlets mainly during winter and spring but some forms flower almost continuously, particularly those sold under the name of 'Payne's Hybrid'.

Origin WA.
Size 1 m high and a spread of 1.5–1.75 m.

Thryptomene saxicola

Climate Temperate and semi-tropical; some forms do not accept repeatedly heavy frosts.
Aspect Full sun to light shade.

Other requirements as *Thryptomene australis.*

Thryptomene australis

Thysanotus patersonii

Fringed Violet, Fringed Lily

FAMILY LILIACEAE

A tuberous herbaceous plant with a tuft of small, short lived basal leaves and a string-like stem that twines around neighbouring plants or trails to about 1 m. It then branches into numerous stemlets supporting a series of slender buds opening to three silky textured lilac-purple oval petals about 2 cm long with finely fringed margins. These appear over a long period during spring and summer.

Origin Mainly in the southern half of Australia, in all but the driest temperate to semi-tropical areas.
Size To about 1 m in length.
Soil Well-composted sand or loam with reasonable drainage.
Climate Temperate to semi-tropical.
Aspect Full to lightly broken sunlight.
Water Moderate to ample, at least during spring.
Feeding Light scattering of complete organic or slow release fertiliser well out from base of stem applied in early spring.

Pruning Remove old flowers.
Propagation Seed.
Possible Problems Usually trouble free.

Thysanotus tuberosa

Fringed Lily,
Fringed Violet
FAMILY LILIACEAE

Herbaceous, tuberous rooted plant with scant grass-like base leaves to 15 cm long, often hidden by other plants. Thin wire-like stems appear in spring, branching and terminating in ,a succession of long oval buds. These open displaying three large oval lavender- to purple-fringed and silky textured petals with thin darker sepals. Flowers last only a few hours but appear from early spring into summer.

Origin All States except Tas. and WA. WA has several similar species including twining *Thysanotus patersonii.*
Size 15–20 cm high and only about 10 cm wide.
Soil Well-composted sand to medium loam with good drainage; happiest in stony soil that provides root anchorage.
Climate Temperate to semi-tropical; moderate frost tolerance.
Aspect Lightly broken sunlight with some protection from other plants.
Water Liberal to moderate; withstands brief dry periods.
Feeding Light scattering of complete organic or slow release fertiliser at least a span out from the plant.
Pruning Old flowers may be removed.
Propagation Seed.
Possible Problems Usually trouble free.

Veronica formosa

FAMILY SCROPHULARIACEAE

A gentle little shrub with slender erect branches with small dark glossy green lance-shaped leaves about 1.5 cm long, arranged in opposite pairs at right angles to the stem and to neighbouring pairs. Soft blue flowers to 1 cm across are carried mainly in small terminal sprays during spring.

Origin Tas., mainly on rocky hillsides.
Size 1–1.5 m high and about 1 m wide.
Soil Moist but reasonably well-drained with compost or rotted leafmould added.
Climate Cool to temperate; frost tolerant.
Aspect In cool moist climates; full sun but lightly broken sunlight otherwise.
Water Liberal; mulch well if soil is inclined to dry out.
Feeding Light dressing of organic or slow release fertiliser in spring.
Pruning Prune back lightly after flowering or moderately during winter to keep growth compact.
Propagation Cuttings.
Possible Problems Usually trouble free, particularly in moderately cool climates.

Thysanotus patersonii

Thysanotus tuberosa

Veronica formosa

Verticordia chrysantha
Golden Feather Flower
FAMILY MYRTACEAE

A small shrub with fine heath-like bright green foliage and, during spring, almost covered with small but profuse densely clustered brilliant yellow flowers. Like most verticordias it is a good long lasting cut flower.

Origin Southern WA.
Size 0.75–1 m high and as wide.
Soil The ideal seems to be a mildly acid gritty loam or mixture of medium to light loam, coarse river sand and compost, preferably with a fairly deep but well-drained clayey subsoil to hold some moisture at this depth. Rockery pockets could be filled with similar soil 20–30 cm deep over a clayey loam.
Climate Temperate to drier semi-tropical.
Aspect Open sunny area away from root competition with larger shrubs or trees.
Water Moderately during winter, in summer only when subsoil area begins to dry out, except for new plants. The aim is to encourage roots to travel deeply to this moisture.
Feeding Not advised.
Pruning Do not prune off old flower heads until new growth commences. It seems best to encourage a period of dormancy after flowering.
Propagation Seed or cuttings planted out at the earliest possible stage (a case of grow your own!).
Possible Problems Difficult to maintain the fairly dry soil conditions needed during summer in eastern Australian coastal areas. Conventional container-grown plants are reluctant to extend their root system after planting out, even though they may make extra growth for a time afterwards.

Verticordia chrysantha

Verticordia cunninghamii

Verticordia cunninghamii
Cunningham's Feather Flower
FAMILY MYRTACEAE

A medium sized shrub with mid-green fine needle-like opposite leaves and large fairly loose clusters of fluffy cream flowers about 1 cm across during late winter and early spring.

Origin NT (Kakadu) and northern sections of Qld and WA.
Size To 2.5 m high and as wide.
Soil Grows mainly on sandy clay and silty loams of flats seasonally flooded during summer.
Climate Tropical to semi-tropical; unlike most other verticordias it is at home in wet humid summers; perhaps a good grafting stock?
Aspect Open or lightly broken sunlight.
Water Liberal during summer, then drought tolerant.
Feeding Tolerance not established.
Pruning After flowering if necessary.
Propagation Seed and possible from cuttings.
Possible Problems Resistance to frosts not yet established.

Verticordia densiflora
FAMILY MYRTACEAE

A small shrub with fine heath-like leaves to about 1 cm or slightly longer (the plant pictured is growing amongst casuarina) and densely packed clusters of flowers about 1.2 cm across. These are usually soft pink with darker pink

Verticordia densiflora

Verticordia grandiflora

centres but may be all pink, purple or occasionally white, flowering from spring to late summer.

Origin Southern WA.
Size 0.6 m high and to 1 m wide.

Requirements as for *Verticordia chrysantha*.

Verticordia grandiflora

FAMILY MYRTACEAE

A shrub very similar to *Verticordia chrysantha*, with leaves slightly longer, flowers slightly larger and turning from gold to bronze as they age. It flowers from late winter to mid-spring.

Origin Southern WA.
Size 0.6–0.75 m high and a similar width.

Requirements as for *V. chrysantha*.

Verticordia insignis

FAMILY MYRTACEAE

A small open shrub with succulent, grey, blunt triangular, overlapping and stem-clasping leaves. Loose terminal clusters of beautifully feathered flowers occur during late winter to mid-spring. These are all pink or creamy white flushed pink with deeper pink centres.

Origin WA, south of Perth from the coast to east of the Darling Range (Avon district).

Verticordia insignis

Size 0.6 m high and nearly 1 m wide.
Soil Occurs naturally in white sand but that suggested for *Verticordia chrysantha* should also be suitable.

Other requirements as for *V. chrysantha*.

Verticordia mulleriana

FAMILY MYRTACEAE

A 'different' erect growing verticordia with stem-clasping, opposite, circular, grey-green leaves about 1.3 cm across and erect spikes of shell-like silky textured flowers which vary in colour from chocolate-bronze to soft purple. These appear from mid-spring to mid summer.

Origin WA, mainly in the warmer Geraldton areas.
Size 1.5–2 m high and as wide.
Soil As suggested for *Verticordia chrysantha* except that this plant accepts a slightly more alkaline soil.
Climate Temperate to semi-tropical; tolerates light frost only.
Aspect At least two-thirds sun.

Other requirements as for *V. chrysantha*.

Viminaria juncea

Viminaria juncea

(syn. *Viminaria dentata*)
Native Broom,
Golden Spray
FAMILY PAPILIONACEAE

Not to be confused with some of the southern European and African brooms which in some of our cooler mountain areas are attractively colourful but becoming invasive. This large shrub appears to have leafless branches as the rather scant 'leaves' or phyllodes, to about 22 cm long, look like thin twiggy bare stems. Growth is gracefully pendulous with cascading spikes of large clear yellow pea flowers from mid-spring into early summer. It is an attractive feature plant near water.

Origin Cool and temperate areas of all States except NT.
Size 3–5 m high and 2–4 m wide.
Soil Adapts to most water retentive soils. Ideal for poorly drained but not stagnant soils.
Climate Cold to warmer temperate; frost tolerant.
Aspect Open sunny or at least half sun.
Water Liberal.
Feeding Moderate dressing of complete fertiliser in spring if necessary to speed growth.
Pruning Only when necessary; resents heavy pruning.
Propagation Scarified seed.
Possible Problems Apart from an occasional flower bud caterpillar it is trouble free.

Viola hederacea

Native Violet
FAMILY VIOLACEAE

A delightful little groundcover with slightly lobed or entire kidney-shaped (orbicular) leaves about 2 cm across, smaller or larger depending on growing conditions and dainty white to mauve-flushed flowers with prominent violet centres on fine erect stems usually a few centimetres above the foliage. An excellent groundcover for not too dry and lightly shaded situations. Invasive, if allowed to run where unwanted, particularly in rockeries, but it is ideal if the right spot is chosen for it. Also suits low containers.

Origin Eastern Australia and Malaysia.
Size Foliage height 5–9 cm mainly depending on moisture, width gradually increasing. Set at 20–30 cm spacing.
Soil Any average garden or bush soil.
Climate Cold to tropical.
Aspect Flowers more freely with at least half to dappled sunlight but also grows in complete shade; accepts occasional footfalls and mowing.
Water Liberal; appreciates moist soil but withstands near complete dryness for short periods.

Verticordia mulleriana

Feeding Rarely necessary; soluble highly nitrogenous type encourage excessive leaf growth; standard complete high phosphorus types promote flowering.
Pruning May be cut with mower set high if excessively leafy.
Propagation Division and runners.
Possible Problems Overexuberance; weed competition while establishing unless well mulched between plants.

Viola hederacea
(Blue Form)
FAMILY VIOLACEAE

A miniature of the species already described with smaller leaves and flowers but the latter are a soft lavender blue with deeper centre. Because of slower and less vigorous growth it is easily smothered by more rampant cover plants therefore is more suited to growing in isolated rockery pockets or delightful as a container plant.

Culture otherwise as *Viola hederaceae*.

Waratah
See *Telopea*

Wattle
See *Acacia*

Westringia fruticosa
Coast Rosemary
FAMILY LAMIACEAE

A very adaptable shrub with lance-shaped leaves about 1–2 cm long in whorls of four, dark green above, grey underneath with grey stems. White to slightly lavender-flushed flowers in the leaf axils occur during most of the year. The overall appearance is a soft misty grey-green excellent for backgrounds as a foil for stronger colours or to create a feeling of distance. Easily brought to dense compact form by occasional trimming. Also use as a grafting stock for some difficult prostantheras.

Origin NSW especially close to seacoast.
Size Exposed to wind and sun, about 1 m high and as wide; in more sheltered aspects 2 m or more high and 3 m across.

Viola hederacea (Blue Form)

Viola hederacea

Westringia fruticosa

Westringia glabra

Climate Cool to tropical; frost and salt spray tolerant.
Aspect More compact in open sunny positions.
Water Liberal to moderate; good drought tolerance.
Feeding Optional; light applications of complete or slow release fertiliser acceptable in spring and summer but rarely necessary.
Pruning When needed; accepts frequent clipping.
Propagation Cuttings.
Possible Problems Extremely trouble free.

Westringea fruticosa 'Morning Light' is a cultivar with cream foliage variegations. It is slower and smaller growing than the species, rarely exceeding 0.6 m high and 1.5–2 m wide.

Westringia glabra

FAMILY LAMIACEAE

In make-up similar to *Westringea fruticosa* but leaves are slightly broader (elliptical), their upper surface comparatively glossy, flowers are a little more profuse but come mainly in spring and are a defined bluish mauve.

Origin Vic., NSW and Qld.
Size 1–2 m high and 1–1.5 m wide.
Soil Adapts to most well-drained soils.
Climate Cooler temperate to tropical; frost tolerant.
Aspect Full sun to half shade.
Water Moderate to liberal but tolerates dry periods.
Feeding Optional light applications of slow release or complete fertiliser preferably in spring.
Pruning Improved by pruning after spring flowering.
Propagation Cuttings.
Possible Problems Usually trouble free.

Westringia longifolia

FAMILY LAMIACEAE

An open shrub with bright green slender linear leaves to about 5 cm long. Outer branches are often cascading and liberally sprinkled with long lipped white flowers. This species more than others is confused with prostanthera but the difference lies in the calyx as detailed in genus notes for Prostanthera.

Origin Qld and NSW.
Size 2.5 m high and as wide.
Soil Adapts to both light sandy and heavy loams with good drainage. All types are improved by addition of compost and a leafy mulch.
Climate Temperate and semi-tropical; frost tolerant.
Aspect Accepts nearly full sun but more valuable because it grows and flowers well under trees if shade is not too dense.
Water Moderate to liberal but accepts dryness well.
Feeding Very little; compost or leafmould are sufficient but slow release or complete organic fertilisers may be given sparingly in spring.
Pruning Moderately after flowers finish.
Propagation Cuttings.
Possible Problems Usually trouble free.

Whalenbergia communis

Australian Bluebell, Grass Bluebell, Tufted Bluebell

FAMILY CAMPANULACEAE

An annual or sometimes perennial plant with a few linear to narrow elliptical leaves mainly towards the base of green wiry branching stems that carry upfacing sky blue to violet, or occasionally white, bell flowers about 3 cm long and nearly as wide at the flared mouth. This is very similar in flower form to the more prostrate and suckering *Whalenbergia gloriosa*, the ACT floral emblem, and in some areas classed as a form of *W. stricta*. Irrespective of exact classification, all are delightful plants.

Origin All Australian States including the dry inland.
Size To about 30 cm high and clumps to 20 cm wide.
Soil Adapt to most well-drained soils.
Climate Cool to semi-tropical; frost tolerant.
Aspect Full to dappled sunlight.
Water Light dressing of complete plant food prior to planting or in late winter.
Pruning Old flower heads removed.
Propagation Seed; perennials by division or cuttings.
Possible Problems Usually trouble free.

Zieria arborescens

Tall Ziera

FAMILY RUTACEAE

A tall fast growing shrub with slender erect usually brown branches, dark green glossy and aromatic leaves divided into three lance-shaped leaflets to 10 cm long and umbels of small starry white flowers about 1 cm across from the leaf axils in spring. Like *Boronia* which it closely resembles, it has four

Westringia longif

Whalenbergia communis

Ziera smithii

petals. But where *Boronia* has eight stamens, *Ziera* has only four.

Origin Tas., Vic. and NSW.
Size To about 4 m high and 3 m wide.
Soil Most reasonably well-drained and well-composted soils.
Climate Cool and temperate; frost tolerant.
Aspect Happiest in fairly shaded areas but will take at least half sun.
Water Liberal to moderate; once established withstands dry periods.
Feeding Very sparingly with slow release or organic plant food in spring.
Pruning May be pruned back as flowers finish or tip prune new growth to thicken foliage.
Propagation Cuttings.
Possible Problems Trouble free, except for an occasional leaf-rolling caterpillar.

Zieria aspalathoides

Heathy Zieria
FAMILY RUTACEAE

A small shrub with erect but closely set branches and small bright green foliage giving a pleasant compact effect. The leaves are divided into three small lance- to salver-shaped leaflets to about 1.5 cm long. During spring they are intermingled with clusters of small pink flowers. It is an attractive rock garden or small feature plant.

Origin Vic., NSW and Qld.
Size 0.5–1 m high and a similar width.
Soil Well-drained mildly acid soil or heavier loams lightened with coarse sand and/or rotted leafmould.
Climate Cooler temperate to semi-tropical; tolerates frost.
Aspect Open sunny position or lightly broken sunlight.
Water Moderate to liberal but reasonable resistance to dry periods.
Feeding As for *Ziera arborescens.*
Pruning May be pruned back by about one-third after flowering.
Propagation Cuttings.
Possible Problems Usually trouble free.

Ziera smithii

FAMILY RUTACEAE

Rounded and attractively compact shrub with dark green, glossy, rather pungent leaves divided into three slender lance-shaped to oblong leaflets about 2 cm long. During early spring it is liberally decked in clusters of small pale pink or white flowers nearly 1 cm across.

Origin Qld and NSW.
Size To 2 m high and as wide.
Soil Most well-drained soils, preferably with compost added.
Climate Cooler temperate to semi-tropical.
Aspect Best in light to moderate shade.
Water Liberal to moderate.
Feeding Sparingly if at all; see *Ziera arborescens.*
Pruning May be pruned lightly after flowering.
Propagation Cuttings.
Possible Problems Usually trouble free.

Ziera aspalathoides

Zieria arborescens

Orchids

Australia is fortunate to have such a wealth of native orchids. Overall, they have a more subtle beauty perhaps than most of the spectacular hybrid exotics that we see in florists' shops but they are nevertheless delightful and fascinating. Another point in their favour is that most of them need only a small growing area, allowing plant lovers with very limited space to enjoy growing a large collection, even in such situations as a reasonably sheltered wall of a balcony, along a fence, in a small fernhouse or under a small tree. Many grow and flower well in situations too shaded for successful growth of most other plants.

The Different Orchids

There are over 700 different Australian orchids, and there are separate works giving coverage of all the species. In this book therefore, only some of the most widely available and easily grown types are covered, partly as an introduction to orchids in general and also to give a general idea of the growing requirements for orchids.

Dendrobium bigibbum

There are two main groups of Australian orchids, each needing different treatment: *epiphytes* which grow naturally on trees or rocks; and *terrestrials* which grow in soil.

Epiphytes usually start their life by germinating in moss or lichens on trees, rocks or where small amounts of organic matter have collected in fissures in bark or croches of tree limbs. They then send their roots over the rocks or tree surfaces to gather nutrients from decomposing microflora such as algae, protozoa, bark flakes, or perhaps in some cases by absorbing free nitrogen from the air. The main point to note about this group of orchids is that although they also need to gather moisture, most prefer their roots uncovered by soil.

This means that all but the few larger growing epiphytes will survive successfully for years on a small slab of treefern fibre or ironbark only about 12 sq cm or, in some cases, no more than half this size. These blocks can be wired with a hook at the back so that at least sixty plants could be supported on a vertical piece of reinforcement mesh or wire netting only 1 m square.

The plants are secured to the fibre by binding them with strips of pantyhose or, less obtrusively, with nylon fishing line, if necessary cushioning roots with a small amount of moss or other fibre to prevent the nylon line from cutting.

Most epiphytes will also grow in pots of compost coarse enough to allow the free access of air. A suitable mix would be cymbidium compost coarsened further by adding up to one-third coarse weathered pinebark or treefern fibre chopped to about walnut size.

Many terrestrial orchids can be grown in any good potting soil or mixtures of bush sand and well-rotted leafmould although it appears that at least some need the presence of 'friendly' soil fungi (mychorriza) to assist their growth. However, this would normally be present in the soil they are purchased in. It is a good idea to be advised by the supplier. Most species can be grown successfully for at least a few years in flower pots or containers only 12–15 cm across.

Cymbidum suave

Epiphytic Orchids

Cymbidum suave

Snake Flower

FAMILY ORCHIDACEAE

A tree dweller that germinates usually in hollows or dead branch butts of hardwood trees; eventually its roots find their way into the decaying heartwood on which it survives, although it has also been found growing in the bark crevices of *Banksia serrulata*. Leaves are deep green, strap-like to about 40 cm long, furrowed down to a strong midrib, ten to twelve from each crown and sheathed at the base in the stubs of older leaves. During spring in the tropics, or early summer further south it carries arching or pendulous racemes 15-25 cm across, green to greenish brown, the yellow lip heavily blotched red, and with a sweet vanilla fragrance.

Origin From northern Qld to southern NSW coast and mountains to 1000 m.
Size Clumps to about 60 cm high and as wide.
Growing Medium Planting in a hollow log filled with composted leafmould, rotting wood and weathered pinebark is the next best thing to heartwood.
Climate Tropical to cooler temperate.
Aspect Dappled shade to full sunlight.
Water Liberal to moderate during summmer.
Feeding Complete soluble plant food or slow release fertiliser during early summer.
Pruning Remove old flower spikes.
Propagation Adventitious growth may form from cracks further down on tree. Otherwise seed is sown onto heat sterilised towelling over a pot of moist sphagnum moss, protected from dust under a glass cover and watered from the base with boiled water; progress is slow.
Possible Problems Once established usually trouble free.

Dendrobium

Dendrobium is a large genus, and most species are easy to establish as they produce clumps of numerous pseudo-bulbs (bulbous or thickened and usually with long stems between rhizome and leaves or in the case of some like *D. teretifolium* and *D. linguiforme*, thickened storage leaves) which provide a store of nutrient and moisture while new roots are re-establishing on divided sections of clumps. As *Dendrobium* they usually grow away from soil contact, these pseudo-bulbs also allow them to survive periods of drought.

Dendrobium bigibbum

Cooktown Orchid

FAMILY ORCHIDACEAE

The floral emblem of Queensland and undoubtedly Australia's most colourful and outstanding orchid. Small clumps of very slender pseudo-bulbs to about 45 cm long, terminate in five to ten lance-shaped leaves 7-10 cm long and arching racemes, to 20 cm long, displaying from five to fifteen showy flowers 3-5 cm across. Size and colour of flowers are variable, usually from pale rosy purple to deep magenta, just occasionally white or near white; flowers mainly from late summer to late spring.

Origin Cooktown through Cape York to Torres Strait Islands, usually on trees, occasionally on rocks.
Size As above.
Growing Medium In moist tropical areas may be bound to trees with a little sphagnum moss or palm fibre covering roots, otherwise in containers of coarse open compost with addition of up to one-third rock or brick crushed to 1-2 cm pieces.
Climate Tropical to semi-tropical; needs glasshouse protection elsewhere. The cold during most temperatde winters causes flower buds to fall.
Aspect Lightly broken sunlight to light shade and shelter from cold wind.
Water Liberal during early to late summer, moderatley dry during autumn and winter.
Feeding Complete soluble fertiliser in early summer.
Pruning Remove old flower spikes.
Propagation Division of clumps.
Possible Problems Cold or too wet during winter in 'borderline' climates.

Dendrobium falcorostrum

Beech Orchid,
Dorrigo Orchid
FAMILY ORCHIDACEAE

This lovely dendrobium forms clumps of pseudo-bulbs 15-27 cm long, slightly grooved, little more than 1 cm thick towards the centre and tapering to either end, terminating in four or five broad lance-shaped dark leathery green leaves 8-10 cm long. In spring short sprays of seven to twelve waxy white flowers, 2-3 cm across, appear with a sweet gently wafting fragrance.

Origin Mountains of southern Qld and northern NSW, rarely less than 1000 m altitude, growing particularly on Arctic Beech, also acacias and treeferns.
Size Clumps protrude 23-30 cm, width depending on age, 15-50 cm.
Growing Medium Happy on the southern side of old stumps, most smooth barked trees, *Banksia serrulata*, treefern fibre or coarse well-aerated compost in terracotta pots or wire baskets.
Climate Temperate to semi-tropical moist mountain or humid coastal areas.
Aspect Prefers dappled or all but very dense shade to full sunlight.

Dendrobium falcorostrum

Water Abundant preferred but they adapt tto moderate; water mostly in spring or early summer.
Feeding Slow release or complete soluble fertiliser after flowering into early summer.
Pruning About flowering time the rhizome may be cut back between old and well-matured pseudo-bulb to encourage regrowth from former.
Propagation Division of clumps.
Possible Problems Dendrobium beetle attacking new growth.

Dendrobium gracilicaule

FAMILY ORCHIDACEAE

Clumps of long thin pseudo-bulbs from 30-60 cm long, sometimes larger under rainforest conditions, terminating in three to six fairly bright green, broad lance-shaped leaves and in spring, racemes, 7-10 cm long, of up to fifteen small shiny flowers ranging from lime green to buff-yellow, sometimes with brown spots and usually fragrant.
Origin East Australian coast from south of Wollongong, NSW, to the Atherton Tablelands in north Qld, mainly on trees with non-shedding bark.
Size Clumps protruding to nearly 1 m and as wide.
Growing Medium As for *Dendroloium falcorostrum*.
Climate Temperate to tropical,

preferably with fairly high humidity.
Aspect As for *D. falcorostrum*.
Water As for *D. falcorostrum*.
Feeding As for *D. falcorostrum*.
Pruning May be back cut as *D. falcorostrum* but needs little encouragement to produce new psuedo-bulbs.
Propagation Division of clumps when all growth has matured.
Possible Problems Usually trouble free.

Dendrobium x *gracillimum*

(syn. *Dendrobium speciosum* var. *gracillimum*)
FAMILY ORCHIDACEAE

A natural hybrid between *Dendrobium gracillicaule* and the Rock Lily or King Orchid, *D. speciosum*. The stout, rather erect and usually brownish pseudo-bulbs, to 60 cm long, terminate in three to five dark green lance-shaped leaves, 10–12 cm long. During spring racemes, 6–10 cm long with five to fifteen dull cream flowers, smaller than those of *D. speciosum*.

Origin South-eastern Qld to central NSW in a few isolated areas — on both rocks and trees.
Size Clumps to about 60 cm high and as wide.
Growing Medium May be grown in the same way as *D. falcorostrum* or on rocks as suggested for *D. speciosum*.
Climate Temperate to semi-tropical.
Aspect Preferably about half but accepts full sun.
Water Liberal to moderate, especially in spring or summer.
Feeding Needs little but grows larger with slow release or complete soluble fertiliser in spring to mid-summer.
Pruning Remove old flower heads; may be back cut as *D. falcorostrum*.
Propagation Division of clumps while growth is dormant.
Possible Problems Usually trouble free.

Dendrobium x gracillimum

Dendrobium kingianum

Dendrobium kingianum

Pink Rock Orchid
FAMILY ORCHIDACEAE

An appealing little dendrobium with great variation in plant form, flowershape and colour, varying according to locality of origin. For example, in the Dorrigo area it usually has comparatively long thin and often purplish pseudo-bulbs to about 30 cm long; the flowers are held more erect, usually coloured deep purple; whereas in the Comboyne areas further south, the pseudo-bulbs are usually only about 5–8 cm long, stout and green, with rounder petalled pink or mauve flowers. There are also white, deep and pale

ndrobium gracilicaule

pink forms, most carrying racemes of up to fourteen flowers in spring. The clumps eventually form mats on cliff faces, or occasionally on fallen logs because plantlets often form at the tip of pseudo-bulbs and eventually fall to start other colonies.

Origin South-eastern Qld and northern NSW.
Size As above.
Growing Medium Fasten or anchor with a few stones on mossy rocks, adding a little leafmould; can also be grown on moist logs, treefern fibre or in pots or baskets of coarse open compost with coarsely crushed brick or sandstone for stability.
Climate Temperate to tropical with some frost protection.
Aspect Lightly broken sunlight to light shade; rarely flowers well in dense shade.
Water As for *D. falcorostrum.*
Feeding As for *D. falcorostrum.*
Pruning As for *D. falcorostrum.*
Propagation From plantlets, division or replanting old leafless pseudo-bulbs.
Possible Problems Dendrobium beetle attacking new growth.

Dendrobium linguiforme

Dendrobium linguiforme

Tongue Orchid, Thumbnail Orchid

FAMILY ORCHIDACEAE

A delightful little orchid which spreads its filigree of creeping stems and foliage decoratively over rock faces. The leaves are about the size and shape of a large soft green almond kernel. A well-established plant looks wonderful during late winter to early spring when massed with 6–12 cm long feather-like racemes of fine petalled white flowers.

'**Origin** From the Burdekin River in Qld to far south coast of NSW.
Size Spreads eventually to about 1 m.
Growing Medium Devise fasteners or use small stones to press plant in good contact with at least slightly weathered (rather than recently cut) rock or on wood or treefern fibre.
Climate Semi-tropical and temperate; preferably in moderately humid atmosphere.
Aspect Dappled to nearly full sunlight for the most flowers, but accepts two-thirds shade. Leaves become purplish bronze in full sun.
Water Appreciates occasional wetting during spring and summer; good drought tolerance.
Feeding As for *Dendrobium teretifolium.*
Pruning Remove old flower spikes. When well established, stem may be severed in several places after flowering to encourage branching.
Propagation Carefully lifting sections and shading until re-established.
Possible Problems Usually trouble free.

Dendrobium speciosum

Rock Lily, King Orchid

FAMILY ORCHIDACEAE

By far the most widely grown Australian orchid; seen in so many temperate to tropical gardens. Makes spreading clumps of stout pseudo-bulbs to 6–7 cm thick at the base, usually curving like a cow's horn and tapering to 20–30 cm long. The pseudo-bulbs terminate in stiff oval to elliptical leaves 20–25 cm long and about 8 cm wide; in early spring, showy spikes 20–50 cm long appear from leaf junctions or terminally, packed with waxy flowers about 2.5 cm across, usually cream ageing to yellow. There is also a deep yellow form with comparatively short pseudo-bulbs.

Origin Coastal eastern Australia from Cape York to Victoria, usually on cliffs or rocky outcroppings, sometimes high in rainforest trees.
Size As above, clumps can spread 1–2 m.
Growing Medium Clumps may be started on rocks, supported by a few well-placed stones (occasionally loosely cemented!), or on logs, in baskets or containers with coarsely crushed rock and leafmould or compost. Do not cover roots with soil.
Climate Tropical to temperate (not severe frost).
Aspect Need at least one-third sun to assure good flowering.
Water Liberal in early summer, otherwise spasmodic.
Feeding Complete water soluble fertiliser acceptable from after flowering to early summer. Do not feed from autumn to flowering times.
Pruning Remove old flower spikes.
Propagation Division of clumps; seed scattered on lightly mossed rocks may eventually germinate. It seems to take at

Dendrobium speciosum

least seven years to flower.
Possible Problems Dendrobium beetle and fungal leaf spot.

Dendrobium speciosum var. *hillii*

FAMILY ORCHIDACEAE

Similar to the species previously described but the pseudo-bulbs are more erect, almost straight, and longer, to about 1 m. Leaves are usually a little larger, and also the flower racemes, up to 60-70 cm, with slightly thinner petalled flowers. They are profuse, giving a soft feathery appearance; usually white and ageing to cream.

Origin Coastal NSW and Qld, more frequently on trees than rocks.
Size To 1 m high and as wide.
Growing Medium Hollow logs or large pots with coarse compost containing pieces of crushed rock or treefern fibre added to keep it open and to give firm anchorage. They also grow on rocks, stumps or logs if given initial support until roots provide anchorage.
Aspect Dappled or about one-third sunlight. They frequently flower well in aspects a little too shaded for the species.
Water As for *Dendroloium speciosum.*
Feeding As for *D. speciosum.*
Propagation As for *D. speciosum.*
Possible Problems As for *D. speciosum.*
Pruning Remove old flower spikes.

Dendrobium teretifolium

(syn. *D. calamifolia*)
Rat's Tail Orchid, Bridal Veil Orchid
FAMILY ORCHIDACEAE

A very dainty orchid when in flower, spreading along branches of casuarinas or paper bark melaleucas by creeping rhizomes. Thin bracket-like branches support pendulous cylindrical fleshy green leaves tapering like rat's tails from 10-50 cm. long. During late winter to spring it drapes a lacy canopy of slender petalled white flowers from a filament of thread-like branches. There are also cream, yellowish and brown-flowered forms.

Origin Coastal northern Qld, almost to Vic. border, usually near to or over rivers or swamps where atmosphere is moist.
Growing Medium Gently but firmly tie to a length of stout casuarina or paper bark limb and keep in a damp lightly shaded position at least until new roots begin anchoring to their new host.
Climate Cooler temperate to tropical with some frost protection.
Aspect Lightly broken sunlight or dappled shade.
Water Spray with water fairly frequently during late spring and summer.
Feeding Not needed in good environment but watering occasionally in spring or early summer with a solution of seaweed extract is beneficial.
Pruning Remove old flower sprays.
Propagation With established plants divisions can sometimes be carefully made just as flowers finish.

Dendrobium teretifolium

Cryptostyllis erecta

Sarcochilus hartmannii

FAMILY ORCHIDACEAE

A delightful little orchid, each erect stem or crown carrying six to eight thick, fleshy, curving and deeply channelled, bright green leaves 10–18 cm long. The reddish tan blotched flower stems arch upwards beyond the leaves, displaying up to fifteen flowers to 3.5 cm across, petals usually crisp white with contrasting reddish blotched centre. Flowers appear mainly during mid- to late spring. Flowers entirely reddish crimson have also been recorded.

Origin South eastern Qld, south to Hunter River in coastal areas. Now almost extinct in original habitats, mainly due to execessive plant collection.

Size Clumps in cultivation rarely more than 20 cm high and about as wide; often larger in wild.

Sarcochilus hartmannii

Soil Grows naturally on rocks or base of trees where roots can reach decomposed leafmould. Does well in pots or baskets of good cymbidium mix or coarsely crushed sandstone and leafmould.
Climate Temperate and semi-tropical with frost protection.
Aspect Flowering and growth is better with at least half or lightly dappled sunlight. Others of the species *Sarcochilus fitzgeraldii* and *S. falcatus* prefer more shade.
Water Keep moderately moist from flowering time to early autumn.
Feeding Complete water soluble food applied once a month from early to late summer will boost growth.
Pruning Remove old flower stems.
Propagating Division of clumps.
Possible Problems Rarely affected.

Terrestrial Orchids

Cryptostyllis erecta

Bonnet Orchid
FAMILY ORCHIDACEAE

Summer flowering usually in loose clumps or small colonies, each plant with one to three elliptical erect and stiff basal leaves 7–10 cm long and about 3 cm wide towards the centre. The flower stem is also erect and smooth for the lower 16–20 cm then terminates in a loose raceme of two to twelve flowers with a prominent broad hood-like labellum 2–2.5 cm long, translucent, flushed green at the tip then veining etched in chocolate to purplish bronze deepening at the base. Petals and sepals slender green and spider-like.

Origin Vic.'s east Gippsland, north into Qld, mainly from mountains down creeks of coastal gullies and in patches of moss on rocky hillsides.
Size as above.
Soil Usually in moist, mossy, well-drained areas with plenty of leafmould.
Climate Cool temperate to semi-tropical; unaffected by frost.
Aspect Lightly broken sunlight to light shade.
Water Liberal to moderate.
Feeding Sparingly if at all in summer with seaweed extract or light sprinkling of organic fertiliser.
Pruning Remove old flower stems.
Propagation Seed.
Possible Problems Rarely affected.

Elythranthera brunonis

(syn. *Glossodia brunonis*)
Purple Enamel Orchid
FAMILY ORCHIDACEAE

A miniature jewel of WA's bushland; it has small tuberous roots, with a solitary lance-shaped hairy leaf and small flowers, 2–4 cm across, with fairly broad, glossy slightly reflexed purple to pale violet petals whose texture is like enamel; the flowers are suspended on fine wiry stems. They appear between late winter and early summer.

Origin Southern WA, particularly Stirling Ranges and surrounding area.
Size 15–30 cm high and 15 cm or so wide.
Soil Grows naturally in fairly well drained composted sand or gritty loam.
Climate Temperate.
Aspect Seems to prefer lightly broken sunlight.
Water Moderate.
Feeding Does not seem to react well to inorganic fertilisers.
Pruning Remove old flower heads.
Propagation Seed.
Possible Problems In cultivation it seems to die out within about two years. Requirements or life cycle not yet fully understood.

Glossodia brunonis

See *Elythranthera brunonis*

Pterostylis reflexa

Small Autumn Greenhood
FAMILY ORCHIDACEAE

One of about 70 species of interesting 'greenhoods'. Like many of the greenhoods, this plant forms a rosette of elliptical basal leaves but in this case they are separate from the plant and usually not present by flowering time. The slender flower stem is on a separate shoot and clasped by three to six slender lance-shaped leaves; it carries a solitary pale green-hooded flower 2–3 cm long with dark green veining and brown tip. These appear in late autumn to early winter.

Origin Mainly eastern Vic. and occasionally in temperate NSW.
Size 2–24 cm high; colonies may extend over several metres.
Soil Occurs both in quick drying granite soil and fairly moist but reasonably drained loam.
Climate Cool temperate and temperate.
Aspect Lightly broken sunlight or light shade.
Water Liberal to moderate.
Feeding Compost or light autumn dressing with complete organic fertiliser.
Pruning Not applicable.
Propagation Seed.
Possible Problems Apart from snail, slug or caterpillar damage are fairly trouble free.

Elythranthera brunonis

Thelymitra ixioides

Spotted Sun Orchid
FAMILY ORCHIDACEAE

One of the loveliest terrestial orchids with a channel led and tapering basal leaf 10–15 cm long clasping a slender erect stem with up to nine pointed oval buds about 2 cm long. These are usually soft blue with a rosy purple flushed centre ray down the outside of each petal. They open during sunny periods to beautiful cobalt or soft lavender blue flowers about 3 cm across with small darker blue flecks on the upper petals. There are also white, pink, violet and mauve forms. The plant has two globular tubers just below its fleshy surface roots.

Origin SA, Tas., Vic., temperate regions of WA and NSW; also New Zealand.
Size Usually about 30 cm high but in some areas to 60 cm; single stemmed rather than a clump.
Soil Prefers lighter loams or composted

Pterostylis reflexa

sand with reasonable drainage.
Climate Temperate.
Aspect Broken to full sunlight (usually closes once completely shaded).
Water Liberal to moderate; fair drought resistance.
Fertiliser Organic, seaweed extract, etc., is safest.
Pruning Remove old flowers.
Propagation Seed.
Possible Problems Usually trouble free.

Thelymitra ixioides

Selecting Native Plants

Shrubs for very Frosty Districts with Hot Summers

Acacias — most species for temperate areas.
Agonis juniperina
Astartia fascicularis
Baeckea (most species).
Banksia ericifolia
Banksia integrifolia
Banksia serrata
Banksia spinulosa
Bauera
Callistemon citrinus
Callistemon viminalis
Calytrix tetragona
Correa species
Crowea exalata
Crowea saligna
Dodonaea
Eremophila most species.
Eriostemon most species
Grevillea all but tropical species.
Hakea bakerana
Hakea eyreana
Hakea salicifolia
Hakea sericea
Hypocalymma angustifolium
Hypocalymma cordifolium
Indigofera australis
Isopogon anemonifolius
Isopogon anethifolius
Kunzea all species
Lambertia all species
Leptospermum
Lomatia
Melaleuca diosmifolia
Melaleuca fulgens
Melaleuca laterita
Melaleuca squarrosa
Melaleuca thymifolia
Micromyrtus ciliata
Olearia all species
Persoonia pinifolius
Peterophile
Phebalium
Pimelea ferruginea
Pimelea ligustrina
Pimelea linifolia
Platylobium formosum
Prostanthera species
Pultenaea flexilis
Pultenaea stipularis
Pultenaea villosa
Ricinocarpus pinifolius
Telopea species
Tempeltonia retusa
Westringia species
Zieria

Windbreak Plantings

Small Trees & Large Shrubs:

Acacia adunca
Acacia baileyana
Acacia longifolia
Acacia saligna
Banksia ericifolia
Banksia integrifolia
Callistemon citrinus
Callistemon salignus
Callistemon viminalis
Grevillea hookeriana
Grevillea Ivanhoe
Hakea salicifolia
Hibiscus tileaceus
Kunzea ambigua
Leptospermum laevigatum
Leptospermum petersonii
Melaleuca armillaris
Melaleuca bracteata
Melaleuca hypericifolia
Pittosporum phillyroides

Tall Trees

Acacia decurrens
Acacia glaucescens
Acacia pycnantha
Eucalyptus most species
Grevillea robusta
Lagunaria patersonii
Melaleuca linariifolia
Melaleuca quinquenervia

Shrubs & Small Trees for Seaside Planting

(Tolerant to salt winds).
Acacia longifolia var Sophorea
Banksia ericifolia
Banksia integrifolia
Banksia grandis
Banksia serratifolia
Calothamnus quadrifidus
Cassia artemisioides
Chamelaucium
Correa alba
Correa reflexa
Crinum pedunculatum
Crowea saligna
Hakea bakerana
Hibbertia scandens
Hibiscus tiliaceus
Lagunaria patersonii
Leptospermum flavescens
Leptospermum laevigatum
Leptospermum scoparium var rotundifolium
Leptospermum squarrosum
Melaleuca armillaris
Melaleuca hypericifolia
Myoporum parvifolium
Olearia
Ricinocarpus pinifolius
Templetonia retusa
Westringia fruticosa

Note that once a salt filtering barrier chosen from some of the larger "frontliners" suggested above is established, a far greater range of native plants can be grown in their shelter, including boronias, calytrix, eriostemons, dillwynias, grevilleas, isopogons, lambertias other melaleucas, phebaliums, pulteneaes etc.

Plants Suitable for very Wet Soil

Actinodium cunninghamii

Ajuga australis
Anigozanthus viridis
Astartea fascicularis
Baeckea linifolia
Banksia robur
Boronia microphylla
Brachycome iberidifolia
Callistemon citrinus
Calytrix tetragona
Celmisia
Crinum flaccidum
Crinum pedunculatum
Dianella caeurelea
Dianella tasmanica
Epacris microphylla
Lomandra longifolia
Melaleuca thymifolia
Mimulus repens
Patersonia occidentalis
Scleranthus
Sowerbaea
Sprengelia
Stylidium
Veronica formosa
Viminaria juncea
Viola hederaceae

Shrubs & Small Trees for Dry Inland Areas

Acacia accola
Acacia aneura
Acacia baileyana
Acacia chinchillensis
Acacia cultriformis
Acacia glaucescens
Acacia lineolata
Acacia murrayana
Acacia podalryiifolia
Acacia rupicola
Acacia saligna
Agonis flexuosa
Alyogne heugelii
Anigozanthus (most species)
Anthocercis
Banksia
Banksia ericifolia
Banksia grandis
Banksia hookerana
Beaufortia squarrosa
Cassia artemisioides
Cassia nemophila
Cassia sturtii
Crotalaria laburnifolia
Dodonaea
Eremophila (all species)
Eucalyptus caesia
Eucalyptus macrocarpa
Eucalyptus torquata
Gossypium all types
Grevillea candelabroides
Grevillea dryandri
Grevillea eriostachys
Grevillea Honey Gem
Grevillea Ivanhoe
Hakea bucculenta
Hakea eyreana
Hakea multilineata
Hakea purpurea
Prostanthera magnifica
Prostanthera striatiflora
Senecio magnificus
Templetonia retusa

GLOSSARY

Acid soil See 'Acid or Alkaline Soil?'.

Alkaline See 'Acid or Alkaline Soil?'.

Anther The pollen-carrying section, normally at the outer end of the stamen.

Awn A bristle-like part or appendage.

Axillary (shoot or flower) Emerging from the axil or junction of leaf and stem or branch.

Biennial A plant normally completing its life in two years.

Basic fertiliser One containing a mixture of nitrogen (N), usually in the form of sulphate of ammonia, phosphorus (P), as superphosphate and usually in the largest percentage, and potassium (K), usually as potassium chloride or in the 'refined' mixtures, potassium sulphate. Generally a sprinkle-on-type dry mixture mixed into the soil before planting.

Bract A modified leaf, in the case of waratahs, some dryandras, etc., actually backing or framing the flower cluster.

Broken sunlight Direct sunlight interrupted by twiggy growth or perhaps sparse eucalyptus foliage to cast a mottling of shadow.

Calyx Sepals forming outer section or backing of the flower, behind the petals and usually persisting when the latter have fallen.

Capsule A dry seed receptacle which opens at maturity, formed from the uniting of two or more modified leaves. Usually shell-like rather than solid and woody (as *Grevillea* rather than *Hakea*).

Coarse sand River sand where particles are irregular and angular, causing them to pack with air spaces between them, as opposed to beach sand with particles worn flat, so that they pack tightly. Even so, for propagation of cuttings, washing river sand is desirable to remove fine particles.

Chlorophyll The green colour pigment of plants which with the aid of light as energy is responsible for photosynthesis. Chlorophyll is a complex molecule combining hydrocarbons with nitrogen and magnesium.

Clay A mixture of micro-fine aluminium silicates, eroded by time to a chemically inert state, plastic when wet, emulsifiable with water only by agitation (to form mud) but otherwise so tightly packed to become impenetrable by air and water, therefore useless as a growing medium until converted to loam by combination with sand and organic matter.

Clay-loam A mixture of clay, sand and organic material with clay in the highest proportion. Very water retentive but a good growing medium if well drained and worked only when just damp.

Dappled sunlight Direct sunlight filtered through light foliage such as tall gum trees; a slightly higher percentage of shade than **lightly broken sunlight**.

Deciduous Shedding, usually of foliage during dormant season, but also applied to flower parts, particularly when the style remains attractive after petals and stamens are shed, as in many Proteaceae. Can also refer to bark.

Digitate Hand shaped.

Elliptical Shaped like an ellipse (narrow oval).

Entire Describing a leaf without divisions, lobes or obvious serrations.

Epigynous Applied to flowers where the attachment of stamens, petals and sepals is above the ovary, such as all of the Myrtaceae family.

Epiphyte A plant that grows on another plant but has its own root system and is not parasitic. Includes many orchids and ferns.

Exserted Projecting, as stamens or pistil from a tubular flower.

Family A group of botanically related genera.

Fungus Plants of a lower order, producing spores, but without chlorophyll for photosynthesis, therefore obtaining their carbohydrates by hosting on green plants (parasites) or by performing the useful role of decomposing spent organic material.

Filament Botanically it is the stalk section of the stamen, supporting the pollen-carrying anther.

Genus A group of species with one or several of the same botanical characteristics that relates them more closely than other members of the same broad family.

Glabrous Smooth, devoid of hairs, usually glossy.

Habitat The area where a plant occurs naturally.

Heavy loam See **clay-loam**, which is the same.

Heel cuttings A mature or near mature side shoot pulled away with downward action or cut so that it retains a small section or 'heel' of the older stem wood.

Herbaceous Normally implies drying down each year to a dormant crown or rhizome but also sometimes to a perennial plant without woody stems.

Hybrid A plant raised from the seed of a flower fertilised by pollen from a different species. Most hybrids only reproduce their kind reliably from cuttings, not seed.

Hypogynous Opposite to **epigynous**, as the ovary is above the point of attachment of other flower parts; as in the Rutaceae family.

Imbricate Overlapping, usually of petals in bud, such as *Eriostemon* as opposed to *Boronia* where petal edges are just touching.

Incised Deeply cut.

Keel The two lower petals of Papilionaceae or 'pea' flowers. These usually protrude forward, united into inverted boat shape approximately at right angles to the base of the standard or large back petal.

Kino Gum of the type exuded from the trunk or limbs of eucalyptus trees.

Leaflets The components ('small leaves') of a pinnate or compound leaf.

Leguminous Plant belonging to the Leguminosae family which embraced all plants with true seed pods, now divided into sub-families, Caesalpinaceae (Cassias, etc.), Mimosaceae (Acacias), Fabaceae or Papilionaceae (all the typical pea flowers).

Lightly broken sunlight Slightly more sun than **broken sunlight**.

Limey Soil containing added or naturally occurring lime; see 'Acid or Alkaline Soil?'

Loam Mixture of clay, sand and decomposed organic matter and perhaps silt in various proportions.

Major elements Nitrogen (N), phosphorus (P) and potassium (K).

Mutant A sport or variation from the parent species, due usually to some factor causing a displacement or re-arrangement of genes. The result is usually termed a cultivar and generally only reproduced vegetatively rather than from seed.

Nitrogenous fertilisers supplying nitrogen such as sulphate of ammonia, urea, dried blood, nitrate of soda etc., but with phosphorus and potassium absent or in comparatively low percentages. A fertiliser that promotes leaf growth rather than flower or fruit.

Ob — prefix indicating inversion; e.g., obovate, with broadest section nearest the tip.

Organic fertilisers Animal or vegetable materials containing one or more of the three major elements, nitrogen, phosphorus or potassium. These are released gradually when soil organisms convert organic to inorganic chemicals, which are then soluble and therefore available to plants.

Ovary Base of flower pistil which after fertilisation enlarges and contains the seed.

Palmate Divided like the palm of the hand, but more usually used when the elongated leaflets are attached at or close to the one base point.

Pedicle Stalk of individual flowers in a cluster.

Peduncle Stem of flower cluster or of a solitary flower.

Petiole A leaf stalk.

Photosynthesis Combination of carbon dioxide and water, in the presence of light and chlorophyll, to form carbohydrates within the plant.

Pistil The female part of a flower containing the ovary at its base and stigma at the outer tip.

Puddle Agitating soil when wet, either by heavy watering, digging or planting. Something to be avoided as it destroys desirable crumbly structure, turning clay-loams into airless brick-like masses. Mulch surface to prevent heavy rain from puddling.

Rhizome An underground stem capable of producing growth shoots.

Scale See 'Pests and Diseases'.

Scarified seed Treatment of seed to break natural dormancy and allow germination. Papilionaceae and Caesalpiniaceae usually respond to abraiding the seed between sheets of sandpaper so that water can enter the seed coat. Alternatively, small seeds of the latter families and Mimosaceae respond to placing them in a cup of boiling water or covering the shallow sown seed with towelling, then evenly pouring a litre or two of boiling water over them. The latter is also effective with some Rutaceae while others of the family need abraiding, then placing in a fabric bag and immersing in a bucket of frequently changed water for ten to twelve days to remove germination-inhibiting substances.

Scree A surface mulch of gravel, crushed stone, shale or similar pebble substance. Formed naturally in alpine areas when freezing causes expansion of moisture in the surface of rocks.

Sessile Attached directly to a branchlet or growth stem without petiole or peduncle.

Slow release fertiliser A mixture of major plant food elements in granules enclosed by a membraneous substance that slows down their diffusion to the soil; they feed gradually over a period of three to nine months.

Soil ball The soil containing the plants roots when removed from its container. Usually done by inverting the latter with fingers supporting soil and tapping container rim on a bench or other raised firm area. Alternatively a plastic bag containing the root ball is cut away. The term originally applied to the ball of soil remaining with the roots of plants dug from open ground and 'balled' in hessian.

Soil structure The physical condition of soil. Desirably a loam that crumbles easily when worked in a just damp state, therefore with adequate spaces between the particles for passage of air and water but sufficiently consolidated to have good firm contact with plant roots.

Stamens The pollen-carrying sections of a flower (filaments and anthers).

Terete Almost cylindrical.

Terminal At the tip of a stem or branch.

Tip cuttings Tip growth taken with 5–10 cm of stem when it is losing its sappiness, usually when it's changing from green to brown; cut fractionally below a leaf junction, and strip off the lower half of foliage.

Vermiculite Particles of mica expanded by furnace heat to form cavities between the flakes, allowing large quantities of water to be held within rather than around the particles. The latter factor then allows maximum air penetration, which with the moisture creates an ideal seed-germinating medium; too soft and spongy for any but soft fleshy cuttings.

Whorl Three or more leaves or flowers radiating in a circle from the one node.

INDEX

The equals sign (=) denotes synonym.